SEPTEMBER 2038

The Global Disaster Squads' Finest Hour

DAVID L. DAUM

PROLOGUE

In September 2038, the world was bracing for the two strongest storms in recorded history. A hurricane in the Atlantic Ocean was moving toward the east coast of North America, and a typhoon in the Indian Ocean was heading toward the Asian subcontinent. The Atlantic storm already had the lowest barometric pressure ever recorded, and the Indian Ocean storm, the second lowest.

The annual number of record weather extremes had been declining for a few years prior to 2038. Still, it had long been predicted that such storms were possible given the right set of circumstances. They came together in the late summer (Northern Hemisphere) of the year 2038. Record temperatures had been recorded throughout the world that year. New lows had been seen in both the Northern and Southern Hemispheres as well as new highs in the north. August of that year found the world with the greatest temperature variation it had ever recorded. Snow was falling over Brisbane, Australia, while temperatures above 50 degrees Celsius had been seen in North America, Africa, Asia, and Europe.

Forest destruction and air pollution had combined in the greenhouse effect to alter the earth's atmosphere. Forest destruction was slowed substantially in the early part of the twenty-first century, and air pollution had been drastically reduced by the year 2038. Still, weather extremes caused by the greenhouse effect continued to occur because it was going to take decades for the carbon dioxide in the atmosphere to fall back to a more neutral concentration.

Weather-related deaths also continued at high levels for two reasons. First, the human population had continued to increase, although at a slower pace than during the previous century, and disperse to every conceivable place on earth, thus making every storm potentially deadly. Second, for a storm to set a new record now meant an extreme condition compared to the average of the nineteenth and twentieth centuries, because trends in temperatures, wind velocities, rainfall, and numbers of storms had been steadily increasing since the late 1990s. The earth was paying the price of humanity's ignorance and arrogance about their home—the earth.

CHAPTER ONE

Late August 2038

"Good morning, what's the latest report?" Director Kristen Verba asked Markus, the newest recruit monitoring the screens in the world headquarters of the Global Disaster Squad.

"Good morning," the youngest member of the Squad answered, "everything looks pretty good overnight. We still have several forest fires going, though. The one in western North America is still not under control, but we expect it to be by 1200 hours local time. There is another fire in Brazil near the mouth of the Amazon and another in northern India. Those are both under control and are expected to be out by evening local times. We had some heavy rains in India earlier that helped get the Asian fire under control.

"The only other situation we have now is the earthquake late yesterday in central Asia. It looks like the epicenter was in southwestern China. It registered 5.8 on the Richter scale and left a few people homeless, but fortunately there were no casualties beyond a couple of people hurt by some breaking glass. As our regional base is close by, our units arrived on the scene moments after we picked up the quake on the monitoring screen. We were able to get medical attention to the few injured right away. We expect a full report on damage later this morning, and I'll get you a copy as soon as it's available.

"Oh, also there was a minor problem with a ship in the South Pacific last night. It started to take on water, and we evacuated the crew and are towing it into harbor now. No one was injured in that incident. I think that is pretty much the story overnight," Markus said.

"Thank you. When do you go off duty?" Kristen Verba asked.

"At 1200 hours, then Natalia takes over for the next shift. Look, here is an update on the quake," he said as information began to come over one of the computer screens that linked anyone watching them to all monitoring devices in the world.

The command station was a medium-sized room with several computers hooked up to sensors around the world. On the wall in front of the computers were three large maps of the earth, each displaying different information simultaneously concerning the world. One map showed weather data in real time from everywhere on the planet. Temperatures were shown in color, barometric pressure in contour lines, wind direction and speed in vectors, and so on. Another map showed hot spots for example, a fire, earthquake, or avalanche; severe weather such as a tornado, hurricane, blizzard; and any other phenomena that needed attention.

The third map showed world data. A person could, for example, find out the population in an area that might be affected by a storm, breaking dam, tsunami, and so forth. One could find out evacuation routes that might be used in an emergency, how many bridges crossed a flooding river, where dams were located, the size of a forest where fire had broken out, natural or man-made firebreaks, and so forth. This map could bring up whichever piece of data might be needed for some given circumstance about any region of the earth.

The room was a whirl of lights and sounds, all spewing forth information about the planet's physical condition. Any earthquake, fire, flood, storm, or any other call for help would automatically come into this room, the control headquarters for the Global Disaster Squad. Workers here would dispatch the nearest Squad unit immediately to the area affected.

Markus continued, "They are finding more people in the rubble. So far, most are still alive, although one entire family of five was found dead in their home. Here's a partial list of other damage coming on line now. We've got several bridges down, three schools, one hospital with minor damage. An earthen dam is leaking, and they are working on that right away to keep it from breaking completely. More later, it says."

"Let's mobilize some reserves in that area to start the rebuilding," the director said. "Start with a level-one call for now until we see the full extent of the damage."

"I'll get right on that," Markus responded.

"I'll be in my office when any more information comes in. How are the tropics, by the way?" she asked, as it was hurricane season in the Northern Hemisphere.

"Pretty calm. We are watching one disturbance off the coast of Africa, but that is about all right now."

"Good. Talk to you later," she said and turned to leave.

She she got to her office, the resident calico cat, Silverton, was lying on her desk, purring as usual. She had been found in Silverton, Colorado, and adopted the Squad's headquarters as her home. She had full run of the headquarters, but would most likely be found on the director's desk. She was really fond of the director, and whenever Kristen was anywhere near, she would run and jump on the director's lap. She really loved Kristen, and the feeling was mutual.

The Squad, as its members called it, had grown out of the militaries of several countries. It was a volunteer organization with members from any given place on the planet. It was mostly young people because of the strenuous activity that was required to be a member of the world's disaster squad. It had started in the early twenty-first century when the prime minister of Sweden challenged the other countries of the world to donate a portion of their military budgets and equipment to the creation of a world disaster unit that would assist in the event of natural disasters. This unit would help in evacuations, assist in medical attention to the injured, help build infrastructure (roads, bridges, schools, hospitals, etc.), and do anything else that was needed. At first few joined her effort to create this body, but then one after another, the developed countries joined the cause.

As the threat of world war continued to diminish after Gorbachev and the first Bush left office in the late twentieth century, it was logical for the world's military to become peace organizations. And simultaneously the threat from natural disasters was increasing due to the destruction of forests and air pollution, which was causing more extreme weather patterns worldwide. The potential for lives lost was increasing along with the growing human population, although the rate of increase was declining as Third World women increasingly took control of their reproductive lives.

To keep ahead of rising natural threats, more and more nations began to move money and equipment from their militaries to the Global Disaster Squad. After all, they had all the machinery needed to rebuild infrastructure that was damaged in natural disasters, and who could build a road faster than a military unit trained and outfitted for such operations? Many nations sent personnel as well as money and supplies. By the year 2038 about half of all former military budgets and equipment had been donated to the Global Disaster Squad.

The Squad was mostly an on-call reserve-type organization, along the lines of the old volunteer fire departments in the United States but on a vastly different scale. There were also groups kept on active duty everywhere in the world for immediate action and in the headquarters. Mostly rescue and medical personnel were on the active duty roster with the construction people on reserve. Of course there were pilots, drivers, and mechanics always on duty to get the others to the disaster areas. And some specialty groups such as the firefighters and underwater units were on duty year round, as they were always needed somewhere. Finally, the support people worked year round as well. The system worked smoothly and responsively.

Kristen had been the director of the Global Disaster Squad for two years now, and she liked her job. It was not an easy job, and sometimes the hours were very long. But it was rewarding work. After she entered her office and petted her cat; she watched the headline news report from around the world, as was her practice each day. She needed to keep current with the world situation. Although most of the world was enjoying the greatest period of peace the planet had ever had, there were still some areas of concern. The political situation was for the most part quite stable except for some ethnic conflicts in remote areas of the world. The Squad was not involved in any politics though. She needed to know about such events in case there was a natural disaster in one of these hot spots; she had to know what the impact would be on the Squad. It had been a while though since one of these political problems had interfered with any activity of the Squad.

Silverton woke up and demanded some attention. Her loud purring could be heard even in the hall as people passed by outside. She jumped up on the director's shoulder and was purring so loud that she almost obscured the news report.

The director did the report on the earthquake and the fires they were fighting. Another part of her daily routine would be to get on the Internet and scan local publications and news stories throughout the world. There was the sports report too. She, like most of the world, always enjoyed a good soccer match. And the World Cup was playing. The first matches had started the day before and her home team from Australia had won. The highlights of the soccer games were just starting.

There was also a message on her e-mail from her old friend Karen, who lived outside of London, congratulating her for her nation's win in the World Cup. It seemed England had won also, and Karen was interested in a small wager. The winner would be flown to the other's home for a week's visit. Kristen sent back a quick reply— *"I'm IN!"*—thinking that she hadn't been to London in several years and it would be just great to see one of her best friends. They had met years earlier while they were both traveling in Bali and had kept in touch ever since.

Karen was quite the world traveler, going to Greece, Turkey, Asia, and Africa. She sometimes traveled with another mutual friend, David. David was a photographer from the United States. She was thinking that she hadn't heard from David in a couple of years and was wondering where he was and what he was doing now days. She made a mental note to ask Karen if she had heard from David and went to scan the World Cup highlights.

"Director," came a call over the speaker. "We just had a severe aftershock in that Asian earthquake."

"I'm on my way," she replied. So much for the soccer game. She hurried along the well-lit halls leading to the control room, thinking about the last earthquake in the world two weeks ago. It also had a severe aftershock that was actually stronger than the first quake. The number of strong earthquakes had been steadily increasing over the last fifty years. The earth seemed to be in a state of convulsion.

"What did it register?" she asked Markus as she entered the control room.

"Six point three," Markus said. "It was stronger than the first one, same as the last big quake. I hope this is not a new trend in earthquakes. It puts the entire rescue squad in extreme danger. Here come the reports from the field." The screen started flashing the latest from the disaster area. He went on, "Some confirmed dead, more injuries, more damage."

"Go to level two on the reserves, we're going to need more help," she said after hearing Markus's report.

"I'm on it now."

"Good. Are the level-one calls there yet?"

"Yes. They arrived right before the aftershock. They checked in when they got there, but I haven't heard from them since then."

"Can you contact them now and see their status, please?"

"Right away." He began contacting the first reserves through the Squad's secure channels.

"Anything?"

"Not yet. Wait, it's coming through now. They are all okay, but they did suffer some damage to some of their equipment. Mostly vehicles. A building collapsed on their trucks as they were pulling into the town. They were caught under the building for a few minutes in their vehicles, but managed to dig their way out fairly quickly. Here are some more casualty figures. They need more blood ASAP. I'm sending that call right now to the medical supply units."

"Good," she replied and watched the data come in from China. Seven strong quakes this year alone in major population areas. And it was only August. That was almost one a month. The Squad was still busy at the last three sites of devastation, rebuilding infrastructure. This was one of the few times the Squad had been so active at so many quake-hit areas at the same time. More information came over the computer screen indicating even more damage than first thought. It was night on the other side of the world, making it more difficult to ascertain the full extent of the damage.

"Director, look at this," Markus said, indicating the map of current conditions. "That fire in the American West has gotten worse. Winds have come up that were not expected, and they are driving the fire in a new direction." It would be crossing ridges toward Lethbridge and Calgary, making it very hard to fight, and threatening residential zones. In addition, it threatened Glacier and Waterton Lakes parks, protected areas where many endangered species of plants and animals lived. The situation had definitely taken a turn for the worse.

"What level are we at in this fire fighting?" the director asked.

"Two."

"Go to four right away; we need to stop it, now!"

"I'm on it now," Markus responded to the director's command.

"I'll be back in my office. Keep me informed right away if anything changes. We don't want this to get completely out of hand."

"Will do. And I'll leave those instructions with Natalia when she relieves me at noon," Markus said as the director went back to her office where she would be meeting with some leaders from around the world early in the afternoon. She entered her office and went to work preparing for her guests.

Markus continued to monitor the situations around the world from the command post of the Squad. His tall body topped with a mop of blond hair was in great contrast to the woman from Brazil with her long black hair and short slender body who arrived in the control room to relieve him of duty. Natalia was from Rio de Janeiro and had been a part of the Squad for over three years now. She was a language expert, and although every member of the Squad had to know at least five languages, she was fluent in at least eight or nine and had basic knowledge of another twenty. Her ability was an innate quality; she could just pick up languages like a good musician can just pick up a tune by ear.

As she walked into the control room a few minutes before noon, the newest information from the earthquake was coming through on the screen.

"Hi, Markus. What's new?" she asked. He updated her on the situations around the planet and the instructions from the director. They talked for a while about current conditions in the problem areas, and then Markus wanted to go and get some rest.

But he wasn't too tired to ask her one more question before heading for his quarters. "So what are you doing later this evening?"

"No real plans. Why?"

"Well, I was sort of hoping you would join me for dinner and a movie later. What do you think?"

"Sure, that sounds like fun. What time? I get off duty at sixteen hundred."

"Say, nineteen hundred. I'll come by your place, if that's all right."

"Sounds great; see you then. Bye." At that he left with a smile on his face, a bounce in his step, feeling very happy. He had wanted to ask her out ever since he first met her. He went back to his room and quickly fell asleep.

Meanwhile Natalia took over the post of monitoring the world's condition. The fire was still blazing out of control in the Rocky Mountains, but the third- and fourth-level reserves were beginning to show up. There had been no more aftershocks in the Asian quake for a few hours, now giving relief efforts a chance to help the injured.

With the major events stable, Natalia went and checked all the screens. The world looked pretty calm elsewhere except for that wave disturbance off the coast of Africa. It was intensifying. There also seemed to be a little weather action in the Indian Ocean due south of the Indian and Bangladesh border region. She made a note of that in her report and returned to the quake screen.

New information was coming into headquarters. Updated casualty figures showed many had been killed in the aftershock; they had been in already damaged buildings trying to rescue still trapped people when the second, more powerful quake had hit. Now even more people were pinned in collapsed buildings. The rescue operation was going slowly, as it was barely first light in the area. She sent the newest data to the director's office as soon as it was available.

The rest of her shift was pretty uneventful with nothing new in any of the world's problem areas. The fire in North America was beginning to come under control again with the efforts of the latest firefighters added to the first groups there, and the quake relief operations were well under way with the arrival of daylight in the region. She was relieved from duty at 1600 hours and went back to her room to send e-mails to friends and get ready for her evening with Markus.

As all Squad members had their own apartments near headquarters, Markus walked the short distance from his home to Natalia's to arrive at a few minutes after 1900 hours.

"Good evening. Come on in," she said, after answering the knock on her front door.

"Hello. Thank you. My, you look great tonight," Markus said, looking her up and down. "I've never seen you before out of uniform. You are very pretty."

"Why, thank you. You look pretty good yourself. Care for a drink before we go?", she asked as she put on some background music.

"Yeah, that would be nice."

"What would you like?"

"Scotch on the rocks, if you have it."

"Would Glenlivet be all right?"

"Perfect."

"Here. Cheers!" she said after handing him his drink. They clinked glasses and sat down to talk. "So you're the grandson of the founder of the Squad, is that right?" she asked him.

"Yea, that's right. My grandmother was the prime minister of Sweden at the time. She was a visionary for world peace and global cooperation. She, Bono, and Oprah won the Nobel Peace Prize for their efforts in starting this outfit.

"Who's Bono?" Natalia interrupted, not asking about Oprah, as almost every woman in the world knew of her. Oprah's work with young women in South Africa was legendary.

"Oh, he was the singer in a rock group called U2 who got heavily involved in promoting certain causes in the late '90s and early this century. He did wonders for Africa getting the major nations to forgive the debt of the very poor countries and getting the world's attention to the AIDS crises in Africa before the vaccine was discovered in the '20s, and he helped a great deal in setting up sustainable businesses for the people there. I met him when I was young when he came to visit my grandmother."

"I know who you mean now; I just forgot his name. We knew him as the singer of U2. Their music is still played today, you know."

"Yeah, I know. I have most of their music at my place. Anyway, to finish about my grandmother, she donated her entire monetary award to the formation of the Global Disaster Squad and a great deal of her time. Soon after she left office, she became the first director of the Squad and led the effort to get the major superpowers to join. She passed away when I was ten years old, and I've wanted to follow in her footsteps ever since. She would spend hours telling me stories of the first years of the Squad and the problems they had in those days. It left me with this urge to be part of the Squad and hopefully add to the effectiveness of our group someday."

"I'm sure you will with your background and determination. Are you from Stockholm?"

"No, the other side of Sweden, Malmö; although we lived in Stockholm for a great deal of my life. And you, you're from Brazil, right? Rio?"

"Yes, from Ipanema, which is one of the areas of Rio. Like Manhattan is a part of New York City. My mother is the head of the natural history museum there in the city. That's what got me interested in natural events. My father is a photographer and explorer. He has gone with my mother on many of her archeological digs throughout the world taking photos of what she finds and her work. They are great team together. What about your mother and father, what do they do?"

"My father was also the prime minister of Sweden at one time early this century. He served two terms and then retired to write books, which he had always wanted to do. My mother was a bioengineer and was part of the effort that found the cure to cancer. Her work was very important to the final breakthrough and development of the vaccine to heal and prevent cancer. She won a Nobel Prize in biology and chemistry for her effort."

"Wow! You do come from some bright people, don't you? There must be a lot of pressure on you to do something important, huh?"

The young Swede blushed. One of his family's traits was modesty. "Not really. My family has always supported all of us kids and anything we wanted to do. I have two sisters; one is a schoolteacher for handicapped children, and the other is a painter. And that is just fine with our parents. They did want us all to follow in the family tradition of politics, but when none of us wanted to do that, they always supported what we did choose to do."

"Mine are like that too," she said, thinking of her parents and how they hold hands any time they were near each other. "So are you ready to eat? I'm getting quite hungry."

"As a matter of fact, I am. Let's go." After putting the glasses in the kitchen, they left for dinner.

They went to a Thai restaurant that specialized in authentic Thai meals from the last century. The food was great, and after their meals they finished with an after-dinner cordial before going to the movie that started in half an hour.

"Have you ever been to Sweden?" Markus asked Natalia as they sipped on their drinks.

"No, but I've been as close as Norway. I went on a cruise in the fjords once when I was younger with my parents and my brother and sister. From there we went to Scotland so my parents could play golf on the oldest courses in the world. Then my mother was attending a conference in London about something, I can't really remember the details. My dad took us kids and went sight-seeing while she was in her seminar. I think my favorite place on that trip was Stonehenge. I was really taken by that place. It was a great family adventure.

"We were lucky; we got to go on many such trips because my mother was always attending these conferences. It was one of the reasons I wanted to join the Squad, to continue traveling around the planet. I have already been in five places since joining the Squad before being transferred here. Have you ever been to Stonehenge?"

"Yes, when I was younger my family traveled to England, and that was one of the sights we visited. How long have you been in the Squad?" he asked her, changing the subject.

"Three years. I joined when I was twenty-three. And you? Haven't you just joined?"

"That's right. I've only been on the Squad for six months now. I joined as soon as I could."

"So you are twenty-three years old?" she asked, knowing that was the youngest age at which a person could join the Squad.

"Yes, that's right. Then I'm three years younger than you. I hope that's all right," he said.

"That's just fine. It's not how old you are physically that counts, but how old you are mentally that's important to me anyway," she replied to his concerns. Sitting across from each other she held his hands on the table, looked longingly into his eyes, and said, "I like you just fine. I've really enjoyed myself this evening so far," she said and then, changing the topic, asked, "What time is it? It must be about time to go for the movie. By the way, what is it we're seeing tonight?"

"*Dancin' Five*", he answered. "It's the newest of the *Dancin'* sequels. It's just a fun movie, I've heard. The first one came out in 2012, and there has been a new one about every seven to eight years ever since. Surely you have heard of them or seen a few of them, haven't you?"

"Oh sure, they have always been one of my favorites. This will be great," she said, as they paid their bill and left for the show.

"Have you seen all of the *Dancin'* movies then?" he asked her as the walked to the movie theater.

"Yes, all of them, over and over. We had *Dancin'* parties all the time when I was in school. We all owned all four of them back then on video-audio. Listen to me talking like I'm so old. Anyway, then we would watch them all starting with the first one, which in many ways is my favorite, and dance to them until dawn. It was always great fun. I bet we did that at least ten times a year. They were the best party VACs you could have, and everyone, no matter where they were from or how old they were, could relate to some part of those movies. Plus you know that most of the proceeds from those movies were always given to preserving wildlife and the cultural differences of primitive peoples. It was truly ahead of its time."

"Yes, I remember my grandmother talking about the first one when it came out. It was one of the things that inspired her to push for the Global Disaster Squad. After seeing the first movie, she really felt for the first time in her life, she told me, that the world was truly one. She had always believed that, but that was the first time she *felt* it. I remember her using the words 'deep down in my heart.' With the people of the world, dancing was the universal constant.

"I still love the way that first movie started with those people in Jamaica skanking to Derik Marley and the Wailers. Then the scene shifts to New York in the world's most glamorous nightclub, and the people are still dancing to the same song. Then to Africa and the same song, and people are still dancing away, then to Thailand: same song, same dancing. It was great." They bought their tickets and entered the movie theater, where they at once picked up a small popcorn.

"Do you like to sit near the front, middle or the back?" he asked her as they went in the theater.

"I like halfway from the middle to the front, if that's okay with you," she replied.

"I was hoping you would say that," he said and took her hand as they strolled down the aisle. They went to the eighth row and sat down just as the movie started.

Walking out of the theater after the movie Natalia said, "That was as great as all the others, I really enjoyed that, didn't you?"

"Yeah, that was truly entertaining. Would you like to stop for something before going back home? I am on duty at 0800, so I don't have to be up too early. What time are you back on duty again?"

"Not until noon again. Yes, I would love to stop, maybe for an ice cream. I've got a little sweet tooth tonight," she answered.

"That does sound good, and I know a great little place that does terrific desserts. The Snowman. Ever been there?"

"No, but it sounds good. Let's go," she said and squeezed his hand as they walked along the sidewalk. They walked for a few more blocks, talking about the movie, until they reached the dessert shop. After receiving their ice cream cones they continued on their way back to Natalia's home. They stopped at a small park on the way back and sat down to finish their ice cream. After they were done they watched the waning moon rise over the darkness of the night, and as a meteor darted through the black sky, they kissed.

CHAPTER TWO

SEPTEMBER 1, 2038

"Good morning, Markus. How was your evening?" Kristen asked as she made her morning rounds. Silverton was riding on her shoulder, as she did from time to time.

"Very good, very good," he said smiling from ear to ear. "Here's the overnight report. The fires are out everywhere in the world at the moment. The quake relief is going on smoothly. There have been some minor aftershocks, but nothing like that second one the day after the first quake. A small earthen dam broke in Australia and did some minor damage to several cattle stations. No one was hurt in that event."

"What about those tropical disturbances?" she asked.

"It looks like the one in the Atlantic will definitely become a hurricane sometime in the next few days," he replied. "Winds are about forty miles an hour now, but it is intensifying all the time. The computer models project that the storm in the Indian Ocean will become a typhoon at about the same time. Movement for both of them is sporadic at the moment, and we can't predict where they might strike land for now. The projections don't all agree, and they carry a high percentage of error right now."

"Well, it sounds like you are taking care of everything for now. I would like a readout of the areas most likely to be affected by the storms before you go off duty today, please."

"Will do. I get off at noon and will drop it by when I get relieved here. Will that be all right?"

"That'll be fine. See you later." Director Verba said and continued her morning rounds of the Squad's headquarters, picking up Silverton on her way out of the room. Kristen Verba was very pleased with the newest recruit to the Squad. Markus was fitting in quite well, and his work performance was excellent in his first six months on the job.

Even though his grandmother had been instrumental in starting the Squad, he had still gone through the rigorous competition to be selected to join the Squad. Then followed one of the most rigorous six-month training period in medical emergency procedures, firefighting, earthquake rescue training, and so forth that anyone could go through. These members who were on full-time for the Global Disaster Squad were the best-trained individuals in the world for these kinds of operations. The reserve levels were also well trained, but not to this extent.

There were also courses in group shock therapy for communities where there were mass casualties and heavy destruction such as the current quake situation in Asia. To become director, Kristen had to undergo even more years of training and on-the-job experience. Kristen was just entering her office when she heard Markus's voice coming over the communicator.

"Director Kristen. Are you there?" Markus asked.

"Yes, go ahead."

"There's been another quake. This one in the Philippines, just north of Zamboanga on southwestern Mindanao Island."

"What did it register?"

He swallowed as he saw the display. "Eight point nine."

"I'm on my way." Kristen put Silverton in her box, and minutes later she was entering the monitoring control room and joined Markus who already had sent the first emergency call out to Squad members covering that sector of the planet. "Any news yet?"

"No, power is out everywhere in the quake area, and our units haven't reached there yet. They are about an hour out right now. I sent them as soon as we pinpointed the epicenter, and they were airborne in just minutes. The tsunami warning went out automatically."

"Good work," the director told the young recruit. He had performed exactly as he was trained to do in this kind of emergency. It was the job of the monitoring personnel on duty to send the initial team into a disaster area. Further calls for additional level backup were the job of the acting director on duty. All actions were to be reviewed by the acting director, and for now and at for at least the next three years, that was Kristen Verba. She had been appointed to a five-year term as director and had already served two years.

"Let's put two more levels on alert right away. Have them standing by for immediate departure at my signal."

"I'm on it now, Director."

Kristen pulled up a seismology screen on the computer and looked at the lines showing the quake. As she watched there were several aftershocks, but none of them were very large for now. Would there be another larger quake as there had been after each of the last several quakes?

She was still absorbed by the charg when Markus's voice broke her reverie. "Director, here's the first report from our people at the scene." They both watched as the words started coming across the computer screen.

"Tidal wave probability: 99 percent," she read from the screen. "Quick, check on the status of evacuations going for all coastal and low-lying regions. Get maximum level personnel on the way to assist with the evacuation and notify everyone in the region of what's happening. You had better call Australia, Indonesia, Thailand, China, and everyone within a five thousand kilometer radius of the quake."

"On it," Markus said and went straight to work. It was time for him to be relieved from duty, and Natalia entered the room right on schedule. As she came in, Kristen had a quick word with her, and Natalia immediately went to work calling and sending warnings to outlying areas of the quake epicenter for evacuation in the likely event of a tidal wave. The three of them were working with extreme intensity. As soon as the initial calls were all made, they checked to make sure no one had been inadvertently left out. Soon they had confirmation that all possible affected areas had received the tsunami warning. They could only sit back, satisfied that all was under control, and watch the screens as more reports were coming in from the first Squad members on the scene.

"Good day," Kristen said to Natalia, when the opportunity allowed for a moment of pleasantries.

"Hello, Director. Looks like you two have been very busy this morning," she replied. "Hi, Markus," she said with a little wink so subtle that only he could see.

"Hi," he responded, with a wink back.

"Well, it looks like we are doing all we can at the moment, so I'm going back to my office" Kirsten said. She got up and then said, "Call me at the slightest change of anything on this situation. I'll be back in about an hour to check on the tidal wave situation.

"It looks like the computer is projecting the first outlying sites will be hit within about three hours. Go ahead and get some additional medical people on the way to the tidal wave areas. Let's have them join the choppers already in the air so they can drop down right afterward and help anyone who needs it. Hopefully they won't be needed and everyone will be evacuated in time. See you two later. By the way, don't forget that report on those two tropical disturbances, Markus," she said as she left the control room.

"I won't", he replied.

Natalia began monitoring the screens for updated information as she took over the desk. Markus stayed on long enough to finish his report on the computer projections of the two tropical storms' most likely paths as they strengthened. When he was done, he left for the director's office with the data. Before leaving though, he smiled at Natalia. "I had a wonderful time last night and was hoping you might want to do something later?"

"I would love to, Markus. I had a great time too. Come by at the same time as last night, and we can decide what to do then."

"Sounds great," he replied, gave her a quick kiss on the neck, and walked away. He went to the director's office and, seeing that she was in a meeting, left the report in her mailbox. Silverton was purring loudly at the director's front door but didn't stop Markus from dropping off the tropical storm update.

The morning had left him a little tired from all of the activity going on in the control room. After returning to his apartment, he fell asleep right away.

"What's the current situation?" the director asked Natalia, as she came back into the control room a short time later.

"I've got communication with our people on the ground in one of the evacuation areas, and they were just about to give a briefing. I'll put them on speaker."

"This is Medchop One. Do you copy, Control?"

"We copy. This is Director Verba; what's the situation?"

"Hello Director. This is Dr. Berg, Southeast Asia medical team. Evacuations are going very smoothly for the most part. Some of the people don't believe they're in danger, and some won't leave without their livestock. We're taking out as many animals as we can as fast as we can, but it doesn't look like we can get all of them before the wave hits. There just isn't enough time to get everything out. Now we are just trying to get everyone to higher ground. Now we are taking off now to be above the area when the wave arrives."

"Understood. Let me know when you have a visual of the wave. Over."

"Will do. We are circling out over the water right now. Over."

"Medchop One, we are showing the tidal wave from a satellite approaching your location in just a few moments. Anything yet? Over."

"Roger, Control. We see it now. It's coming in very fast. Oh my God! I've never seen anything like this before in my life, it's huge! The outer islands are gone under a wall of water. Completely submerged! I'm afraid if there were anyone left on the islands, our presence isn't going to help them. It was an unbelievable sight.

"Mist is shooting up into the sky now as the water hits the main island. We are being bounced around by the turbulence and are pulling up to avoid losing control. Wait! Control, there's another wave right behind the first one. Do you copy? Over."

"We copy. Is it as big as the first?" Over."

"Doesn't appear to be, but it doesn't really matter as the first one took out everything anyway. We are following it up the coast now. It's still hitting with quite a punch, does not appear to be losing any of its power. Over."

"Keep with it. What else do you see? Over."

"We can't see much right now. Everything is underwater along the coast. The destruction to the plant life is pretty severe though. We'll need the tree planters here right away when this is cleaned up.

"The wave is approaching a fishing village now. We don't see anyone on the streets anywhere. No sign of life at all. The wharf is underwater ... the buildings ... the town itself is gone from view. Shit! Pardon my French, but the whole frigging town is gone. Every building—gone! Nothing! Nothing left at all! We're going to need a lot of help here, Director! Over."

"We copy, Medchop One. Stay with it. We already put out a level-five alert right after we knew a tsunami was inevitable. Over." This was one of the worst natural disasters the world had seen in a long time, maybe even worse than the 2004 Indonesian tsunami.

"We're with it, Control. It's moving out to open water again. There are some smaller islands ahead, and then it's on to Indonesia and Australia in our direction. Have you heard anything from the north yet? Toward Thailand and Vietnam? Over."

"Nothing yet. We're switching to them now. Over."

"We copy. Will stand by. Out."

Kristen turned to Natalia. "Connect us with the other Medchops."

"Connected," Natalia replied almost as fast as the director had asked for the connection.

"This is Control. Do you copy? Over."

"This is Medchop Four. We copy. We've been listening to your conversation with the units to the south. We don't show anything here yet at all. Where does the satellite show the wave relative to us? Over."

"We show it about two hundred kilometers to the south of your current position. Does the evacuation look complete yet? Over."

"Fairly close. There are a few stragglers here and there, but they are the exception and not the rule. There is a herd of elephants that we have been herding toward higher ground with the chopper, and I think they are out of danger now. They have lit some fires along the coast to scare the animals out of the forest and onto higher ground and it seems to be working pretty well. We can see herds of animals, flocks of birds and so forth headed out of the coastal regions and into the foothills from our location. We're about as ready as we can be. Over."

"Thanks for the report. Stay with us, and report back as soon as you see anything. Out for now."

"Roger, Control. Out."

The director and Natalia continued to monitor the situation, talking to each of the Medchops in the air around the region in danger from the tsunami. One by one, they reported on the damage as the wave slammed into one landmass after another. The loss to human and animal life was turning out to be minimal, but the loss of forests and plant life was going to be staggering. Millions of square kilometers were being wiped out in a single catastrophic incident that was destroying some of the most beautiful areas of the world. This was a truly sad day for life on the planet.

"Control, this is Medchop Four. Do you copy? Over."

"We copy. Go ahead. Over."

"We can see the wave now. We are northeast of Ho Chi Minh City about halfway to Nha Trang, Veitnam. It's heading for the coast right underneath us and should hit land in less than a minute. We have a camera on board and are switching it to Control headquarters so you can observe the wave from here too. Here it comes ... Wow! Water is shooting up half a kilometer into the sky!

"It hit with incredible power. Too bad we couldn't harness that energy for constructive purposes and not destructive like it is. It's moving up the coast. The damage to infrastructure is pretty absolute right along the coast. Further inland, its not as bad, but still there will be help needed. The wave has washed up some of the rivers, reversing the flow of water temporarily and wiping out some bridges in the area. We'll stay with it. Over."

"We copy. Out." They continued to listen to the reports coming into the control room and sent out the call for appropriate help as needed. This would be taxing the Squad to its limit in some zones, and reserves from other parts of the world might have to be transferred right away to assist in the rebuilding efforts. Natalia was charting the wave's movement throughout Southeast Asia and had the computers working on the total territory that would end up being affected by the tidal wave. It looked like locations as far away as Hong Kong and the coast of the Sumatra would feel some of the wave's fury, but not to the extent of what was happening here in the Western Pacific region. The tsunami was mostly heading west with little activity to the east of the Philippines.

Markus came into the control room after his short nap and joined the women. "I saw the news at my home and came to see if you needed any help. How's it going in here?" he asked Natalia.

"You wouldn't believe what's been going on!" She updated him on the situation around the quake region and showed him the pictures of the destruction coming in now from the entire sector. The tsunami was moving up the coast toward Taiwan from Hong

Kong now, and a smaller wave was headed out toward the islands of the South Pacific. The entire Pacific basin was still on full alert, and coastal sections throughout the area were being evacuated.

"How many casualties?" he asked. "Do we have any numbers yet?"

"No," Kristen replied. "It's hard to tell. Many districts are completely cut off from the outside, and we aren't able to get anyone in until the water subsides. Also there were two waves, and we have to wait for the second one to hit and then recede also. At first they were right behind each other, but now there is about an hour delay between them. We're hoping that no one returns to the coast prematurely and gets caught in the second one."

"Wow! This has been quite a destructive two weeks we've gone through," Natalia said as the impact of the director's words began to sink in.

"It certainly has," the director answered. "It certainly has."

"Look, more updates," Natalia said pointing to another screen. The situation in the Philippines was bad. It had started out as such a good day for them as their soccer team had won their first match in the World Cup by a decisive margin, and now major portions of several of the islands were laid to waste. Many of the people had been out of the buildings celebrating the victory in the streets when the quake hit. The tsunami hit moments later, catching a lot of the people outside. Manila was saved from a direct hit by the island of Mindoro, but the wraparound wave still was pretty powerful. The quake had been on the western edge of the Zamboanga Peninsula, which sent most of the energy westward, although a weaker wave went out in all directions.

"I've got a meeting that I need to go to now, but if any thing new happens, call me immediately," Kristen said as she left the control room and headed back to her office.

"Markus, are you going to stay here all afternoon?" Natalia asked, after things had calmed down for a moment in the control room. They were watching the tidal wave move toward its next target.

"Why? Do you want me to leave?" he asked back teasingly.

"No, not at all. I enjoy being with you, it's just that you didn't rest very long, and you have to work again tomorrow. I worry about you, that's all," she replied, studying him with a concerned look.

"Thanks. I appreciate your concern, but I'll be all right. I was thinking of getting a few more hours sleep anyway in a while. It's just that this is so—so new to me that I don't want to miss anything that happens. I—"

"I know just what you mean, I was the same way when I was first here too. Couldn't get enough of it, but you'll get over that," she said, interrupting him. They both gave a little laugh and continued watching the screens as the wave smashed into the coastal regions of the Pacific Rim. The wave was losing some of its punch the further it traveled from the epicenter. This was good news to places like Japan and the coast of China. The evacuation of the locations more distant from the quake had been very complete as there was much more time to deal the situation. Positions on the western Americas from Alaska to Chile were starting their evacuations of low coastal regions preparing for the wave to hit them sometime later today or even early tomorrow. Hawaii was already ready as were most of the other islands throughout the Pacific.

"I think that I'll leave now for that nap," Markus said.

"Okay. So I'll see you later? Same time?" she asked.

"Sounds good to me. Bye," he replied and left the control room to return to his quarters and get some well-needed rest.

Kristen entered the control room with the heads of state she was leading on the tour. She explained that this was the room where the entire world was monitored for any natural calamity. Then she asked Natalia, "How's it going in here?"

"The wave has just hit Japan and South Korea," Natalia answered. "Here's the satellite picture of the tidal wave on this screen, and on this one over here is the picture from the Medchops over southern Japan. They're right over the land area as the wave hits to be able to drop down and assist as needed immediately."

The dignitaries moved about the room looking at all the equipment and asking questions as they checked out the control room of the Squad.

"Thank you for your report. I'll be back before you get off duty to get a full update," the director said as she ushered the heads of state out to continue on their tour.

"It'll be ready," Natalia replied.

"Quite a responsible young person you have in there," one of the dignitaries said, after they were out of the control room.

"Yes, she is. And the good news here is that everyone in the Squad is like that, or they don't last very long at all. We have to keep morale high, as they sometimes have to work for days without barely any rest. So people who don't respond well are gone in a hurry. It doesn't happen very often; most people who can't make it are weeded out very early in the training process. Our training is the most comprehensive in the world for this kind of operation. We are the best," the director stated with abundant pride.

"I can see that," the minister responded. "You all can be very proud of yourselves and your organization."

"Thank you," the director said. She finished the tour with them, and after walking the heads of state to their waiting vehicles, she went back to her office. Silverton was purring on the desk and there was a message from Karen. She thought they should go to Queensland after she won, and no, she hadn't heard from David in about a year. But she expected he would show up any time now, as was his custom.

Karen was of Polish decent and would have been a princess in years gone by when there was still royalty in Poland, so she always signed her letters *Karen, your Polish Princess*. Kristen looked forward to seeing Karen; it had been a few years now since they last visited, and she didn't really care where she saw her, although London would be the best.

"Hi, Markus, come on in. I'm almost ready. Help yourself to a Scotch," Natalia said, speaking from her bedroom, still getting ready. Markus closed the door and turned toward the living room.

"Thanks, I will. What can I get for you?" he asked.

"There is a bottle of Merlot on the kitchen counter. The opener is in the top drawer near the wine, and the glasses are right above in the cabinet. Half a glass would be great." She finished preparing for the evening and came downstairs in a black dress that hugged her curves, with diamond earrings hanging almost to her tanned bare shoulders. Markus looked up as she descended the stairs from her bedroom and knew she'd seen him react to her beauty.

"Wow, you look great!" he exclaimed, not even caring that he couldn't come up with anything better. Who could, after a sight like this?

"Thanks, sweetheart. You look pretty good yourself." She put her arms around the tall handsome Swede with his blond hair neatly combed straight back and gave him a long, deep kiss. "Wanna just stay here?" she asked after coming up for air. He started kissing her again, nodding slowly.

CHAPTER THREE

September 3, 2038

"Good morning, Markus. How are you today?" Kristen asked as she started her daily rounds.

"Fine, very fine, and you?" he replied.

"A little tired, but otherwise good. These last few days have been taxing, and I didn't sleep that well last night. What's the overnight report?"

"Well, there's good news and bad news. First the good news: there have been no more aftershocks at either of the earthquake sites in almost thirty-six hours. The cleanup of the tidal wave is going very smoothly, and there haven't been any bodies found since yesterday, with only a few people still missing. All the major fires are out, although the temperatures are reaching record highs in some of the more fire-prone areas, especially western North America and eastern South America.

"Now the bad news: the storm in the central Atlantic has become a full-fledged hurricane and is tracking toward the Lesser Antilles. Winds are near a hundred and forty kilometers per hour already. It really intensified overnight and the pressure is near a record low, so it probably hasn't stopped strengthening. It is moving toward the west-northwest very slowly, soaking up heat from the ocean.

"The other tropical disturbance we're tracking in the Indian Ocean is near hurricane strength and is projected to become a full-fledged typhoon later today or by tomorrow at the latest. That one is barely moving at all right now, and its pressure is also still going down. The record high temperatures around the Northern Hemisphere may account for some of these storms' development. Well, that's the overnight report."

"Thanks. Keep me posted on the storms" she said with a worried frown. "I don't like how low those pressures are getting for this early in their development. Can you get me a report on storms over the last, say, fifty years and send that to my office screen, please?"

"It'll be there before you are," he replied and got up and punched the appropriate commands into the one of the computers in the control room. Moments later the information was coming up on the director's screen in her office.

"Thanks. I'll be back in a few hours," she said and left with a very bad feeling about these two storms. Back in her office she picked up Silverton and petted her on her lap, while she started going over the last fifty years worth of data on tropical storms. Very few had ever had such low pressures early in their development as the two storms now had. Of the seven with such low pressures in the last fifty years, two moved into colder water and faded completely. One was near land when it formed and moved onto land before it could develop any more and dissipated. Another became a strong storm but never hit land. The other three all set records for their time for the lowest pressures recorded to date—and all three hit land, causing a great deal of destruction. This was indeed a bad indication for these two new storms.

Kristen Verba studied the information in front of her and grew concerned. There were serious concentrations of populations within striking distance of both these developing storms, and the impact could be catastrophic. She had always been able to trust her feelings on such things, and this time she didn't like how she felt.

She was getting up from her chair when Markus's voice came over the intercom. "Director, I think you'd better come to the control room right away. There's been another quake."

"I'm on my way." Moments later she was entering the control room, and Markus immediately showed her the data he had on the latest quake. Japan. Five point eight on the Richter scale. Fortunately Japan was built for quakes, as it experienced them quite frequently. "We are checking probability for tsunami now."

"When did it start?"

"About two minutes before I called you. It started out with a few small tremors then really kicked in. See, here's the data on this screen here." The Squad's headquarters were hooked up directly with the world earthquake-monitoring center located in Golden, Colorado.

"Thanks," she said and studied the information for a moment.

"Aftershock! And big," Markus exclaimed as the seismograph started moving radically over the paper. "Six point nine," he said as the machine began to calm down. He had already notified the Squad in that area, and they were in route. He quickly updated them on the current data and turned to the director for further instructions.

"Wave probability?" she asked.

"Sixty-nine percent," he stated. He didn't need to be told to start notifying all affected areas, and in just moments everyone in the region who might be hit by a tidal wave knew about it.

The director then had him put out a level-three call for the epicenter area and a level-four alert for the tidal wave basin region. She would wait until the wave was detected before calling them into the location, but wanted them on call. It would not take long to determine if a wave was going to happen. Even though the probability was high, whether a wave would be generated depended on how the quake had moved along fracture lines. The computer was programmed to predict not knowing that information, and quakes that occurred near water always carried a greater probability for tsunamis.

"I think we've lucked out. There seems to be a wave forming, but it's not very large. Here's the data now. Yes, it will only reach about three meters. It can still do damage in some places, but it will be nothing like a few days ago after that Philippine quake. I'll put out the alert to affected zones," he said after the computer showed which places in the region would be hit by the wave.

"Good work, Markus. Put a level-two alert out for the affected territories just in case they're needed. The computer still has room for error, and any given area might have something else going on that could enhance the wave," she directed him. This was standard procedure for such an event.

More information was coming from Japan. The quake had occurred on the northern island of Hokkaido, near Sapporo. The aftershock had done very little additional damage. Most buildings were still intact, and although there were quite a few minor injuries being reported, mostly from breaking glass, casualties were at a minimum. This was in great contrast to the last two quakes, where many were killed, especially in the aftershocks.

"Here's more data on the wave. Also we're picking it up on the weather satellite monitoring this part of the globe. The computer predicted this wave almost exactly. See?" Markus showed Kristen the latest report.

"That's good. It looks like the newest software is doing the job it was designed to do," she replied after studying the data on the screen in front of her.

"Good morning," Natalia said, coming in thirty minutes ahead of her shift time. "Another quake?" she asked, after a look at the screens. Not waiting for an answer she asked further, "Where this time?"

"Japan. Six point nine. Good morning," Markus said.

"Good morning, Natalia," Kristen said. "Well, as long as you're here early, would you mind updating the data on those two storms in the Northern Hemisphere that we're tracking?"

"I'm on it now," she replied and went straight to work on her task. The Atlantic storm was still drifting very slowly in a generally western direction, intensifying at a steady pace. The Indian Ocean storm, now southeast of Sri Lanka, was basically stationary for now and was nearing hurricane strength. Computers still showed it becoming a full-fledged typhoon within twenty-four hours. With no clear indication on where it was headed for the moment, the Indian Ocean storm was considered a threat to the entire region.

Tropical storm Nisha, as it had been named by the global weather system, was located just north of the equator almost directly south of the Indian and Bangladesh border. And the Atlantic storm, which had been named Hurricane Colin, was about halfway between the west coast of Africa and the Leeward Islands of the Caribbean, moving slowly toward the west. When Natalia had the latest data on the two storms, she gave a summary of the information to the Director.

"Thanks, Natalia," Kristen said after she received the figures. "Doesn't look so good, does it?" she stated more than asked, still looking looked over the report.

"Those are both pretty low pressures for this early in their development," Natalia responded. "Anything else, Director?"

"No, nothing special. Just take over the routine responsibilities now. It looks like the world lucked out for the most part on this one quake so far. It could have been a lot worse. It could have turned into another wave situation like the last quake in the Philippines. Well, I've got another meeting I am required to attend, but as always, keep me posted if anything changes." She looked toward her office; mustn't forget to pick up her notes. "See you two later. I'll be back in a couple of hours. Bye," she said as she headed for the door.

"Bye," they replied almost in unison. Upon entering her office Kristen found her cat sleeping on her desk as usual and purring, also as usual. An e-mail message from Karen revealed that David, their mutual friend, had showed up at her place in London and promised to contact Kristen soon. He was from Colorado, where the headquarters were located, and said he was returning there soon. He indicated that he would look her up there when he returned sometime in the next month or so.

He had been in Gabon for the last year photographing the lowland gorillas and forest elephants. Before that he had been in the central Democratic Republic of the Congo, where he had been living with the people there as part of a year-long sabbatical. Kristen sent a message back and quickly returned to the task at hand for her next group of meetings.

"Let me show you what's been going on," Markus said and updated Natalia on the global situation and then turned over the duties of monitoring the world for natural disasters.

"Thanks for a great night," Natalia told Markus as he was preparing to go and get some rest. "Would you like to come over to my place again tonight?"

"That would be great—or you could come and see where I live if you'd like."

"That'd be great. Give me directions. What do you want to do tonight?" she asked, as he wrote out his address and directions from her place.

"Dinner maybe. I'm sure we'll think of something," he said, giving her hand a little squeeze. They kissed, and he left to get some rest as she took over the controls. The rest of her shift was relatively quiet with the tidal wave turning out to be quite tame in most places. There were a few stretches where it did some minor damage, and a couple of small boats were beached here and there, but for the most part it was just like a very high tide. Only some low-lying areas near coastlines had to be evacuated, and even then the people leaving didn't have to go very far inland to avoid the wave. In fact, most just watched from higher ground.

Kristen came back two more times during Natalia's shift for updates on the quake, the wave, and the two tropical disturbances. "What does the computer project for the Indian Ocean storm?" she asked on her second trip to the control room. "Same as before?"

"Pretty much," Natalia replied. "The Indian storm looks to be a full typhoon very soon. It is already tomorrow there, and is nearing dawn. It's expected that when the heat of the day hits the storm, it will intensify past tropical storm status to a full typhoon."

"Any idea where either of them might be headed yet?"

"Not on the Indian one. It's still just moving about in a tight little circle. And the Atlantic one is still moving only slightly, although it has changed to a more northwesterly course that would take it ashore somewhere in southeastern North America if it stays on this path. It's still too early to say, though, as these storms have a way of changing course all the time.

"Here are the probability sheets on likely landfalls for both storms. As you can see, the Indian Ocean one has a huge error margin posted with each prediction, and the Atlantic one isn't much better."

"Thanks. Keep me posted, okay?"

"As always."

Kristen went back to her office to study the latest reports on the two storms. Silverton came back in through the cat door as she heard the Kristen walk into the room, jumped right on the desk, and laid down on the papers the director was reading. Kristen set the cat in her lap and rubbed her ears. Still nothing solid to go on as to where the storms might hit land.

The Squad had put out all the normal statements on both of the storms to the general public, and although they reported the barometric pressure in those reports, she knew most people didn't comprehend the seriousness of the situation. If these storms continued on their most likely path of development, there was a good chance that both of them would reach record low pressures and record wind speeds. This of course, depending on where they went ashore, might cause record damage and loss of lives as both were poised to hit very populous regions of the planet.

It was, however, too early to panic such a large portion of the world, so the Squad was only issuing very strong statements about the need to follow the development of these storms as the days went by and to begin basic preparation in possible target regions. This was about all they could do until they had some more hard statistics on the two tropical disturbances. Years of experience gave Kristen a feeling about these two storms. And it was not a very good feeling. She had almost a premonition that things were going to get a lot worse before they got better. And then there were the earthquakes to contend with at the same time. This was sure one major time to be with the Squad.

"Come on in," Markus yelled to the sound of the doorbell ringing as he hurried down the flight of stairs into the hallway leading to his front door. He threw open the door. Natalia was standing there as beautiful as ever. Her red skirt and white blouse went perfectly with her black hair blowing gentle in the breeze. He put his arms around her and drew her close to him in a strong embrace. They kissed for over two minutes straight before stopping. Then, looking into each other's eyes, they fell to kissing again.

"Can I come in?" she asked after a while.

"Sure, of course," he answered, pulling her in to the room.

"Nice place you have here."

"Thank you. Come on in. Would you the tour?"

"Please. Is this your parents?" she inquired looking at a picture on the wall.

"Yes, when they were first married. It was taken on their honeymoon in Bali," he said and began showing her the house.

"Are you as hungry as I am?" he said when they finally reached the kitchen.

"Yes I'm starved. Let's get something to eat. Do you like Mexican?" she asked him.

"Love it, and I know a where we can get some great food, and they have the best margaritas in the city."

"Then what are we waiting for?" she asked as they headed for the door.

"Let me put my cats out, and then let's go," he answered. Moments later, he put his arm around her as they left for the best Mexican food in town. Outside his building, they caught a taxi and traveled to the Mexican restaurant. After a brief wait they were seated in a booth along the back wall of the restaurant.

"This is a great drink," Natalia stated after sipping on her margarita.

"Wait till your second one, and then try and talk about it," he replied. They ordered food and talked about the events of the day. Those storms were the most urgent concern of the Squad right now, which they picked up from the director. Kristen's constant asking about the status of the storms made both of them very apprehensive about them. Perhaps out of a sense of foreboding, they spontaneously began kissing right in the booth.

"Are you finished with your plate, madam?" the waitperson asked, interrupting their embrace with a somewhat mischievous smile on their face.

"Yes I am," she replied, mumbling as she was caught up in the moment of romance.

"I'm through, too," Markus added to the waitperson. "I'm so glad I met you, Natalia. You make me feel so wonderful being with you."

"I feel the exact same way. This has truly been a great week. Isn't it great that our shifts were back to back? Otherwise we might never have met." She gazed at him, and he lost himself in her limpid brown eyes. They just sat and looked at each other for a long time. After perhaps thirty minutes the bill was brought to the table. They paid and, after leaving, walked slowly down the street holding hands headed back to his place.

"Would you like a nightcap?" he asked her once they were inside.

"Please." They both had the next day off and ended up talking for most of the night before falling asleep on the floor shortly before dawn. His two cats curled up with them.

CHAPTER FOUR

September 5, 2038

"Good morning, Markus. Have a good day off yesterday?" Kristen asked as she entered the control room on her morning rounds.

"Yes, it was great. Natalia and I went into the national park and saw many animals. Laid around in the sun and otherwise did nothing. Saw three new species of birds too," he answered.

"I didn't know you were a bird-watcher."

"Both of us are. Natalia has over a thousand birds on her life list, and I have about eight hundred. She's traveled a lot more than I have in places where there are more birds, like South America."

"Well, sometime we must compare notes. Even though I don't have much time for it right now, I also am a birder. I have about fifteen hundred in my life list. Most of them I saw when I was younger and still living at home in Australia and Southeast Asia. My mother taught me how to see birds where others never saw anything. Anyway, the overnight report, please."

"We would enjoy that a lot," he responded and then continued with the overnight report. "Here's the latest on the two storms: Both are full-fledged hurricanes and still intensifying. Hurricane Colin is still moving very slowly to the west-northwest on a course that will take it into the Leeward Islands. Winds are over 160 kilometers per hour now, and the pressure is still dropping. Typhoon Nisha has winds near 140 kilometers per hour and has started a very slow drift to the north. Pressures also continue to drop in this storm, too.

"There are still too many variables to predict landfall locations for either storm. Although the computer has given us some locations and probabilities associated with each of those sites. There have been only very minor aftershocks in the quake locations and cleanup and repair are under way in all places affected by the quakes and tidal waves.

"There is a new forest fire to report this morning, caused by lightning, on the island of Corsica in the Mediterranean. It is still small, and firefighting crews have been working since local midnight to contain it. We expect to have it out within six to eight hours from now.

"And there was also a train accident on the Trans-Siberian route, which killed nine and injured over fifty. That could have been a lot worse as a bridge collapsed, but only one car was on the bridge when it went down. A few moments later, and the entire train would have gone down, and there were more than a thousand people on it. There are reports of several weak bridges on that route. Many of them have been in service for a hundred years or more without being upgraded. That's about all there is to report this morning."

"Thank you," Kristen said. "Get some engineers to do a survey of the rest of the bridges on that route that are still questionable and have them send me a report as soon as they're finished. Also I need a full report on the quake and tidal wave cleanup. It needs to include the total number of Squad members involved, amount of equipment used, estimated rebuilding costs and so forth. I'm meeting with some of the major contributor nation states In a few days to ask for additional funds, since we've had so many costly incidents so far this fiscal year. Our reserves are down to dangerous levels, and if we get very many more major problems, we're going to run out of money. And these two storms make me very nervous."

"I'll get right on it, Director," he replied and sent the call out for the bridge report and then started on the report she'd requested. "I'll get this to you as soon as it's ready. Also if anything changes on the two storms, I'll let you know right away."

"Thanks, I'll be in my office," she answered with a little smile of appreciation for his effort. She went back to her office, followed by Silverton. The cat had gone on morning rounds, but decided today to follow the director instead of riding on her shoulder. Kristen was thinking about the upcoming meetings on the financial situation of the Squad. Money was pretty tight right now, and while the major contributors were always generous in their donations and usually sympathetic to the Squad, finances made for very tedious work and the part of her job she liked the least. But it came with the territory, and she always did a superb job detailing the expenses of the Squad down to the last little change.

The meeting would last about two days and would take up a great deal of her time while the visitors were here. She was glad that she had such outstanding staffers working with her at the Squad. If she had to be out of touch for a while, she didn't have to worry about everything falling apart. But then they wouldn't be here if they weren't competent. That was the way it was in the Squad.

The numbers starting coming in on her computer screen from Markus on the current expenditure estimates of the unfolding natural catastrophes. She had a project for Natalia when she came in for her shift at noon. She needed a cost evaluation prediction on the two current storms if either of them were to hit land, along with the probabilities that they would hit land. Then tomorrow the director would work on the likelihood of additional problems this fiscal year. She certainly had her work cut out for her.

"Director are you there?" Natalia asked over the intercom after taking over her shift from Markus at noon.

"Yes, go ahead."

"I had a message to give you a call when I arrived for my shift."

"Yes, good afternoon. I have a project I need you to do right away." Kristen went on to explain what she needed. "Also if you would leave me a report on the two storms at the end of your time in the control room, I would appreciate it."

"You've got it, Director," Natalia replied with her usual enthusiasm.

Kristen came by the control room on her afternoon rounds with Silverton following behind her, as was apparently her new habit, to check in with Natalia, who was hard at work on her project.

"Hi, Director. I was just about to send this over to you." She tore off the computer forecasts of the potential expenses of the two weather disturbances.

"Thanks. How does it look?"

"Could be bad, real bad. If Hurricane Colin continues on its present course, it will affect most of the islands of the Caribbean and most likely Florida and the entire Gulf Coast region, too. It has sufficient strength already to cross over Florida and reform in the Gulf without much loss of energy. We also have noticed that water temperatures in the storm's projected path are very high for this time of year, which should maintain or increase the storm's potential intensity. If it moves further north, the entire East Coast as far as Newfoundland is at risk, although water temperatures are lower on that route, and the storm would be less likely to reach extreme levels. But that might even be more costly as those stretches are even more densely populated.

"As far as Nisha goes, the potential for disaster is even higher. Its present course would take it into some of the most densely populated zones in the entire world. Eastern India and Bangladesh are straight in its current path."

"This is not good news," Kristen said with a worried look on her face. She continued, "We should put out a preliminary bulletin on both storms right away to all possible landfall areas and a general shipping advisory for affected shipping lanes. Also let's make sure we have evacuation plans in place for all possible targets. By the way, how is the blood plasma supply in these areas?"

"Fairly good in North America, but we could use more. There's not much in Asia: a lot of the reserves were used in the quake and tidal wave aftermath. We have put out a call for additional donations in Asia already, and we're starting to get response on that request already. Next I'll get one out for North America right away. Anything else?"

"Not at the moment, but I'll let you know. Thanks for your help."

"No worries, mate. It's me job," she responded in an exaggerated Australian accent, trying to make the director feel more at ease.

"Obrigada," Kristen responded. "By the way, did Brazil win last night in the soccer match?"

"Why yes, we did. We beat Cameroon, defending World Cup champions as you might remember, three to two. It was quite a match with us winning in the last minute. We were tied with only fifty-eight seconds to go in regulation time when we got three shots on goal in an eighteen-second flurry, with the third shot bouncing off the leg of their goalie and in for the score. With only forty seconds left, we got the ball back and ran out the clock to win. And your team?"

"We won too. It was a fairly easy match for us. We played Bulgaria, and you might have heard that they had some injuries in their last game, which they won, but their star striker didn't play. So it wasn't a true test of their skills. But a win is a win, so we advanced to the quarterfinals. I called home this morning to talk to my family about the game, and they were all very excited.

"Anyway I've got to go now, I've got some more meetings to attend. Talk to you later."

"Okay, see you later," Natalia responded, watching Kristen's feline friend follow her out, and went back to monitoring the world for disasters.

The rest of her shift was pretty uneventful, and she was relieved at a few minutes to four. She dropped by the latest update on the two tropical storms to the director who was still in her meeting. On her way out of the complex, she ran into an old friend of hers from the days at the academy.

"Susan. Is that you?"

"Natalia! I didn't know you were stationed here. I thought you were still on assignment in Europe. How are you?" she asked as they gave each other a big hug and a kiss to each cheek.

"I'm great. I was transferred here a while ago. I've got a good position here in the control room, and I've got a new boyfriend. What more could a girl ask for? And you, what are you doing here?"

"I've just been transferred here from Asia. I'm going to be working on upgrading all the computers and monitoring devices here at the headquarters, and it's a two-year assignment. It's so good to see you—and tell me about this new boyfriend of yours. Did you meet him here, or did you know him before?"

"No, I met him here. He's the grandson of the founder of the Squad—"

"Is his name Markus?" Susan asked, interrupting her friend.

"Yes it is. Do you know him?"

"I met him on a training exercise. But I don't really know him. I just remember that he said his grandmother had helped start the Squad back in the early part of this century. He seemed very nice, and he is real good-looking, as I recall."

"Yeah, he is very good-looking and one of the nicest guys I've ever known. He takes very special care of me. I feel this might turn into something, I care for him a lot. How about you, dating anyone?"

"I was, but we're not seeing each other anymore. We just had different lifetime goals. We got along great and always had a good time together, but we could never agree on the long-term stuff. He wanted to get married right now and start a family. But that would mean I would have a hard time staying with the Squad, and I'm not ready to give that up yet. Yes, I want a family someday, but I'm way too young to start now. I've still got some things that I want to do before that stage of my life. I've been given a great job opportunity here with the Squad to enhance my technological skills with this new assignment. I'm in charge of the entire upgrade project, and I wasn't ready to walk away from that. So we broke up. Anyway, when do I get to meet Markus?"

"Well I'm going over to his home later and we were going out to dinner at this new Mexican restaurant that we really like. Why don't you meet us there? I'm sure he won't mind. Here's the address and how to get there. How about 1900 hours?" she said as she wrote the address and directions down on a scrap of paper she found in one of her pockets. They parted, and Natalia went to her quarters, where she quickly changed. She hurried over to Markus's place where he was waiting for her. She knocked on the door and entered.

"Sweetheart—" She started to tell him about her friend who would be joining them for dinner, but before she could say any more, he planted a big kiss on her soft lips and pulled her to the sofa where they fell backward onto it. He continued hugging her, and she didn't mind at all.

"Sweetheart, we have an old friend of mine meeting us for dinner in twenty minutes. I hope that's okay with you," she finally said when they came up for air from their kiss.

"That sounds great, darling," he answered, kissing her again. "Then maybe we should go," he added. "I'll call and change the reservations to three people instead of two," he said and went to make the call. When he came back, she told him about her friend and that he had met her before in a training exercise before.

"What's her last name?" he asked Natalia, trying to remember the person.

"Karosel. She's from Florida."

"Oh yes, I do recall her. Yeah, we were in a special rescue-training program in Asia. I didn't really get to know her very well; we were so busy, we didn't have much time for anything, let alone getting to talk to each other. Well, this should be fun, but we'd better go, or we'll be late." With that they were out the door.

When they arrived at the restaurant and went in to meet Susan, she was waiting by the hostess stand.

Natalia made the introductions. "Susan, this is my boyfriend, Markus. Markus, please meet Susan." They exchanged greetings and talked about when they had met before.

After ordering their meal and some special gold margaritas, they started talking about the Squad.

"So you both work in the control room, is that right?" Susan asked.

"Yup, and that's where we met. Natalia relieves my shift in the monitoring department. And … well, one thing led to another and here we are. It's been great," he added with his big smile radiating across the table.

"Well, I'm happy for both of you," Susan said. "You two certainly seem very happy together."

"We are," Natalia added. "Here's a toast to friends," she said after the drinks arrived. They all clinked their glasses and took a sip of their cocktail.

"It's sure been quite a couple of months for earthquakes lately," Susan said after they had been chatting a while about old times in the academy. "I was in Asia when that one hit and then had the larger aftershock. That was something else. It was the strongest quake I've ever been in, and it was quite scary. I heard about that one in the Philippines too. I was on leave in Florida when that one happened, and that tidal wave—wow! I saw the pictures of it on the news when I was home. Couldn't believe the force of that thing. We were pretty lucky that more lives weren't lost."

"That's for sure, it could have been real bad," Markus responded. "But the evacuations went well, although the planet did lose a lot of forest area. It will take quite a while to replant the destroyed region. By the way," he said, turning to Natalia, "how are those two storms doing?"

"They were getting stronger all during my shift. Kristen had me do some cost analysis on the damage potential of both of them, and it could be very costly. Both in forests and infrastructure. We should be able to keep human and animal casualties to a minimum with proper evacuations, but the forests and buildings could take a real beating. The winds were reaching class five levels when I left. But both storms were still not really heading anywhere very fast. Of course that gives them time to grow even stronger. It's going to be a tough couple of weeks coming up. Kristen is really worried about it."

"Yeah, they both intensified some on my shift. I've got a bad feeling about these cyclones. Real bad."

"Me too," Natalia chimed in. "Anyway, here's dinner. Let's hope we're both wrong, and they fizzle out before they ever hit any land." Changing the subject, she said, "Boy, am I hungry. I didn't eat all day. This looks very tasty."

They continued with small talk for the rest of their meal and afterward stopped to get some ice cream on their way back to their homes. They were sitting at a small table outside the ice cream parlor as a slight breeze cooled the evening air.

"Well this has been quite a night," Susan said as they finished their dessert. "I enjoyed myself very much. Thank you both for a lovely evening I'm glad I ran into you. Next time it's my treat, okay?" she said.

"Sounds good to me," Markus responded. "I never turn down a free meal."

"Thanks again, Markus, for dinner," Susan said. "It was great. Bye, you two, I've got to get some rest. I'm still a little bit jet-lagged. And tomorrow is my first day on the job. I'm meeting the director first thing in the morning, and I want to be fresh and alert. Bye," Susan said again and left Natalia and Markus holding hands in front of the ice cream parlor. They waved good-bye and started back to his place.

"Nice friend you have there," Markus said on their way back.

"Yea, she's a great person. You'll get to like her a lot and she is always ready to party. But very hard-working when she's on duty. You know, she is the top graduate in computer science the Squad has ever had. That's why she is in charge of upgrading the entire computer and monitoring system of the Squad worldwide. She'll be here for two years at the headquarters working on this setup. She's quite a woman. She and I have known each other for about four years now. I met her before I joined the Squad while touring Squad facilities when I was getting ready to apply for training. We usually keep in touch, although this last year we lost track of each other. I'm glad she has been transferred here.

"Anyway let's get to bed. Tomorrow is going to be a long day, I bet, and you have to be on duty by 1000 hours," she said with a sweet little smile on her face, saying that she didn't have sleep on her mind exactly yet. He squeezed her hand and they hurried off to his place.

CHAPTER FIVE

September 6, 2038

"Good morning," the director said, meeting the newest person assigned to the headquarters complex. "You must be Susan. I'm Kristen Verba. Very glad to meet you. Come on in."

"Yes I am, Susan Karosel. Nice to meet you, too. I've heard a lot about you, and it has all been good."

"Please sit down. Would you like some coffee? I've got a fresh batch just finished."

"Yes, please, that would be great." The two women chatted for a few minutes over their coffee about Susan's upcoming assignment at the control center, and then Kristen gave her a tour of headquarters.

"Good morning, Markus," Kristen said as their tour took them into the control room. "I hear you have already met Susan, our new computer expert here, and you know that she will be renovating our computer system over the next two years."

"Yes. Good morning to both of you."

Susan said, "By the way, thanks again for that delicious meal last night, it was truly a culinary delight."

"You're most welcome. Glad you enjoyed it," he replied and turning to Kristen he stated, "Here's the morning report."

"Thank you."

"Earthquake status remains static, with no new aftershocks. Cleanup and rebuilding is well under way, and we now have an estimate of about three more months before everything is back to pre-quake conditions. The fire in Corsica is under control and should be out later today. We do have a new fire in Mongolia. Started last midnight local time by lightning. It's in a very remote forest area and was picked up by the crew of the space station first before any ground confirmation. We have units on the way—in fact, they are showing that they're already at location," he said after checking a screen to the left of his chair.

Continuing, he announced, "a small tornado hit a housing development in southern Spain, in a coastal village near Cartagena. Twenty people are known dead, and another eighteen are still missing. We have had rescue personnel on duty since about five minutes after it happened. Some of our Squad members were on a training exercise just outside the village when the twister hit.

"And speaking of storms, typhoon Nisha, the Indian Ocean storm, has winds over two hundred kilometers per hour, central pressures have dropped to under nine hundred millibars, and sea swells are beginning to interfere with eastern India and Bangladesh," he said for Susan's benefit since the Director obviously knew where the cyclones were located, "and is still intensifying. Look at these figures from the last computer update," he said and continued, "The storm is interfering with shipping seven hundred kilometers away already."

"Better put out a severe shipping lane advisory immediately. Any clue where it's headed?" Kristen inquired.

"Here's the latest computer projection: Calcutta."

"Fifteen million people live in the urban area alone, with another hundred million in the surrounding regions. Oh boy! This isn't good. What's the error percentage on this prediction?"

"Thirty eight percent. It's starting to look like that will be the landfall for the eye of the typhoon. But check out the size of the storm. We have hurricane-strength winds out nearly one hundred kilometers from the eye in all directions now, and that area seems to be still increasing. I put out that shipping advisory first thing on my shift this morning.

"Also I sent a special advisory to the regions where there was a fifty percent chance of landfall or greater. That included most of India north of Madras, all of Bangladesh, and western Burma. Also there's high potential for severe flooding all the way into Nepal, Bhutan and even southern China. Deforestation in those regions has made flooding a major problem. And even though the reforestation project has gone well, the trees are still fairly young.

"The storm is on a northerly course switching from slightly northwest to slightly northeast. The center is now near eighty-seven degrees east and ten degrees north or about seven hundred kilometers due east of the northern tip of Sri Lanka."

"Sounds like you have given all the advisories that need to be out by now. What's its speed at the present time?"

"About twelve kilometers an hour."

"And the Atlantic hurricane?"

"Colin is centered near fifty-one degrees west and thirteen degrees north. It's not as strong as the Indian storm at the moment, but the computer projects it to become much more powerful because of the warmer waters it is traveling through. If that is the case, we're looking at new records for both wind speed and low pressure. It is tracking west-northwest at about fourteen knots at the moment. Hurricane-force winds extend about ninety kilometers from the center, at the center of a tropical disturbance about six hundred kilometers across. This storm is huge already. I have sent all the preliminary advisories and watches to all places with a fifty percent chance of being struck by Colin already and the same shipping alert for the affected water routes. Anything else you want out now?"

"No, looks like you've got it under control. If anything changes ..."

"It'll be in your hands before the ink is dry," he declared before she could finish her sentence.

"That's what I like to hear, Markus. I'll be back before the end of your shift for another update. I'm giving Susan a tour of the complex, and we're not quite finished yet. See you later. Bye."

"Later on. Bye, Susan. Welcome aboard."

Kristen and Susan left the control center and after finishing the tour they ended up in the office where Susan would be working.

Kristen said, "So here is your office. You have access to all the computers at headquarters from this room. Make yourself at home, and call me if you need anything. And again, welcome aboard."

"Thanks, this will be perfect," Susan said, and as Kristen left, she started unpacking the equipment that had already been delivered to her office.

Kristen stopped by her desk to see if she had any messages, and having none decided to go back to the control room to see how the hurricanes were developing.

"I was just about to send you the latest on the storms," Markus said as the director entered the room. "Here's the up-to-the-minute numbers on them," he said and pointed at the screen in front of her. She began to study it.

"They're still intensifying and still on the same basic course as this morning," Markus stated as she read the report.

"I see. Has anything changed other than their strength?" she asked him.

"Not really, except they're both moving a bit slower. Other than that, things are about the same as when you were here earlier. Should we put out further alerts or watches?"

"Maybe later today when we know more about their paths, but since they have slowed down a little, we have some more time. If anything changes though, we need to know immediately," Kristen stated emphatically. "I want you to leave very explicit orders for the following shifts that any change in intensity, speed, or course is to be given to me immediately, anytime day or night. I'll have my pager with me at all times, so if I'm not here, they can still reach me right away."

"I'll leave it in the daily log special alert slot and leave a hand-written notice on the screen also. Here's Natalia now. We can let her know right away. Good morning," Markus said as Natalia came into the control room, and then he updated her on the situation and the director's request.

"I'll pass it on to the next shift after me personally," she responded. "And I'll have each shift pass it verbally and sign the log sheet that they understand the directive," she added.

"Sounds good," Kristen said and left the room to return to her office.

"The director gave Susan a tour of headquarters this morning, and she came through here," Markus said to Natalia after they were alone. "I think she has the old control room as her office so she can access the system without interfering with ongoing operations. At least that is what I heard was going to happen. The custodial staff told me they got the word to spruce up that room by today. So I assumed that is where the director assigned her. Makes the most sense anyway."

"That's probably correct; after all, where else would she go? All the other offices are occupied most of the time." They discussed the current situation with the hurricanes and typhoons and what the director expected if anything changed.

"Anything else going on in the world that I need to know about?", Natalia asked. Markus went on to update her on the rest of the world's disaster situation. After he was done, he left, turning over the command to Natalia, and went to see Susan in her new office.

"May I come in?" Markus asked, sticking his head through the partly open door.

"Of course. Come in. Well, what do you think? This is my new home for the next couple of years. I think I can get to really like this old room. This was the old control room from what I understand, right?"

"That's correct. And it looks like you have a new friend here," he said seeing Silverton lying right in the middle of her desk sleeping despite the cleaning and confusion going on all around her.

"Yeah, she just came in here a bit ago and made herself right at home."

"I see you're making yourself right at home in here too. Anything I can help you with?"

"No, I don't think—well, maybe there is. I think I would like this cabinet and table to switch places so that I can get into these drawers while still sitting at the computer. Would you mind helping me move them around?"

"No problem," he said and started moving the table out of the way so he could push the file cabinet into place. Then, putting the table in place, he asked, "Anything else?"

"Not at the moment. If I think of something more, I'll let you know. By the way, I'm going to have a small group of people over this coming Friday and would really like for you and Natalia to join us. They're some of the people I've known from other assignments who are stationed here, as well as some people I've just met. Plus a friend of mine is visiting here for a few days, and I want her to meet everybody I'll be hanging with for the next two years. I like having parties, so this is a good excuse to have one—you know, so that all these people can meet. Do you think you two will be able to make it?"

"Sure, we'll be there. Wouldn't miss it. But for now I think I need to go home and get some rest. If you need anything, just call. See you later."

"Thanks, Markus. You've been very helpful. See you later. Bye," she said and then returned to the task of preparing her new office for the start of her work.

"How's the organizing going?" Kristen asked, looking in on Susan's new domain.

"Great," she replied. "I think I'll be ready by tomorrow with any luck. Doesn't look like this room has been used in a while. Has it?"

"No, we moved out of here several years ago and have used it only in emergencies for backup since then. I'm on my way to check the control room for the newest update on those two storms. It could get really bad in a few days if they continue to grow and keep their present headings. It's a foreboding situation developing out there in those warm tropical waters. I'll keep you abreast of any news," the director said and left rather abruptly, as if her own words had motivated her to see what was happening right away concerning the two hurricanes.

Susan mumbled a good-bye to her as she left.

"How's it going?" Kristen asked Natalia as she came into the control room. "Anything new on those two storms?"

"Nothing new on their direction, but both of them are reaching near-record low pressures. Look at this, Nisha is now at 875 millibars, and Colin is at 878." She pointed to a screen that showed the two storms' pressure histories, which clearly indicated that one, if not both of the storms, would set new low-pressure records within two or maybe three days at the most.

"I see what you mean. That Indian Ocean storm is just pretty much sitting still for now"

"Yes. And the Atlantic one is drifting slightly west-northwest, and it has slowed to only three kilometers per hour.

"What about the winds? How strong are they now, and what does the computer say about them in the future?"

"The winds have increased to 220 kilometers per hour in both of them, which makes them category-four storms already. We expect them to get stronger before they get weaker. The computer predicts the winds could reach three hundred kilometers per hour if the present trends continue. The storms are reaching very dangerous levels. Both will be cat fives soon and could be that strong when they make landfall.

"I think a level-three or maybe even level-four alert would be prudent at this time for the possible affected regions. We have a level-two alert out now for a very broad area of both storms' probable paths. It's hard to get too excited when we can't accurately predict where the hurricanes are headed. We don't want to put the general population into a panic—or even worse, get them not believing us because we're too vague."

"True, but a level three is proper at this time. With winds that high and projected to go even higher before these two dissipate, we need to get that alert out now."

"Consider it done, Director. Anything else?"

"Not at the moment that I can think of. Oh wait, there is one thing. My sister is arriving here for a brief visit next Tuesday, and I want to have a few people over to my place to meet her while she's here and would like to have you and Markus join us if you are free to do so," she stated.

"I'm sure we would be able to join you. Thanks for the invite. Can we bring anything?"

"No, it's being catered, and they will take care of everything. Thanks for the offer though. I'll be in touch about the time; she was vague about just when she comes in that afternoon. I wanted to invite you now so that you would be able to come.

"Joan is my oldest sister. She's a shuttle pilot for the airlines and works for the most part on the Asia to North America passenger space runs. She's going to be changing her normal route to add the newest spaceport in southeastern Africa when it comes on line next month. So she's taking a brief holiday and visiting some of our family and friends. I hope she can stay for a few days, but we've got a rather far-flung family to visit, so I don't know. Anyway, it will be great to see her even if it's only for a little while.

"I'll be back before the end of your shift for an update. See you in a bit," she said and left for her office where the never-ending pile of administrative work awaited her.

"Okay, see you later." Natalia began another round of routine checks on the world and finding nothing new at this time spent an hour working on the probabilities of where these two storms might make landfall. The cyclone in the Indian Ocean was barely moving, so making an accurate prediction was close to impossible with the data on hand. But anywhere it might strike from Madras northward to Bangladesh and around to Burma or even further south to southern Thailand were all densely populated areas. The potential for disaster was tremendous.

"Anything new?" Kristen asked right away as she entered the control room.

"No. Nisha has pretty much become stationary, only moving slightly in a very tight circle. See, here's its path over the last four hours, and it's right back where it started four hours ago. As far as Colin is concerned, it is still moving west-northwest at about eight kilometers per hour. Fairly slow movement. Winds in both storms remain strong, very strong. Both are category-five storms now with sustained winds of 260 klicks. Seas in both areas are very rough, and I've issued the standard shipping advisories as required. Storm surges are reaching out hundreds of kilometers already. Latest projections are hard to read for Nisha; when the storm moves southward in its tight circle, the computer lessens the odds of landfall and then increases them again when the storm moves north. So, not much to say on the Indian Ocean storm.

"The Atlantic storm is moving so slowly that the potential landfall predictions include a large area. It still looks like the eastern United States is the most likely target from Florida to Virginia. Also the Leeward Islands are in danger, as is Puerto Rico, the Dominican Republic, Haiti, Cuba, and the Bahamas. Jamaica could be hit too, if the storm were to move even slightly more westward. I sent out standard hurricane watches for all these regions with an extra comment as to the severity of the storm. Some lowland areas have already begun to move livestock to higher ground. Small craft have been warned to seek shelter, because no matter where this storm goes, the seas are going to be very rough in all directions for hundreds of kilometers. So the first preparations are under way, and all areas are getting ready for major evacuations whenever we give the call."

"Good work," Kristen replied. There was such an awesome responsibility here in cases like this. Give the call too late, and millions were at risk of dying. Give the call too early, and commerce was disrupted, lives thrown into panic—and then if the storm missed them completely, it was all done for nothing, and the next call for an evacuation would be taken less seriously. The perfect balance had to be achieved, and even though the computers could give a fairly decent guess, the final call always came down to a gut thing. You just had to do it.

"Look," Natalia said, "a more northerly movement in Nisha. It didn't make the circle back to the south that time. Let's see where the computer predicts now …" She drifted off into programming the proper sequence of inputs to achieve the desired results. Then she looked up. "Calcutta!"

"Well, put out the normal alerts to that area, but let's keep watching it before we put out any more warnings. It might change direction again, since it's not really moving fast enough to establish any kind of momentum".

"You got it. I agree, it's a little too soon to get really excited about this movement."

"I'll be back in a little while. Just page me if anything changes. I'm not going to be in my office. See you shortly."

Kristen left the control room with Silverton right behind her. The calico had come into the control room while the director was in there, but no one had seen her until she jumped down from the top of a warm computer when Director Verba left the room.

The rest of Natalia's shift was uneventful, and she was glad of the reprieve. Paul Johnson was scheduled to relieve her at 1600 hours, but he was half an hour early. She updated him on the current situations and orders from the director, and when they were done he took over the post early.

Paul was from Canada near the border of Alberta and British Columbia in a small mining town on the Alberta side of the border. He had been with the Squad for almost four years now and specialized in weather phenomena. So he had a keen interest in what was going on right now with these two storms.

"So what's your gut feeling on these cyclones?" Natalia asked, as she was about to leave the control room.

"It doesn't look good. Both storms are passing over record or near record sea surface temperatures, and Hurricane Colin has a strong high pressure system developing to the north of it, which will tend to keep the storm in warm waters and probably turn the storm's track to an even more westerly path. That would take it through or near many of the islands of the Caribbean and into Florida. Colin easily has the strength to cross Florida and pass into the Gulf of Mexico. If that high-pressure area wasn't developing where it is, there would be a good probability that the storm would turn north and move into colder waters and eventually rain itself out over the ocean, never hitting land. But unfortunately that doesn't appear to be happening.

"Nisha is almost certain to hit land also, since it's moving into the Bay of Bengal with potential landfalls on all sides. And as you updated me, Calcutta is the target for now. So overall, my gut feeling is that this is going to bad, real bad, before it's over," he stated with a chill in his voice. "I'll relay the information to Kristen about the high pressure area right now, as I'm sure she'll be interested."

"Thanks for your insight, even though it doesn't sound too promising for those areas. Hopefully we can keep losses to a minimum with enough advanced warnings and major evacuation."

"Well that's what we're here for," he said, in a lighter tone.

"That's for sure," she replied. "Anyway, see you later. Do you relieve me tomorrow?"

"Yes, so I'll see you then. I will be on this shift for the next few weeks, while the usual person is on holiday. Bye."

"Bye." She left the control room and headed over to see Markus. He was expecting her later, but she wanted to surprise him. When she got to his place, he was working on some plants in his small attached greenhouse with the door open. She slipped in behind him and gave him a kiss on his bare tanned back. He turned around and in a very affectionate embrace they glided to the greenhouse floor.

"Wow! That was so beautiful," Markus said as he held her in his arms on the warm greenhouse floor. "You are so gorgeous. I love you so much."

"And I you." They lay there for another half-hour, each enjoying the comforts of the other's arms.

"So what do you want to do tonight?" Markus asked after a while.

"I feel like dancing tonight," she replied. "I don't know why for sure, maybe because I'm so happy with you," and she gave him another kiss.

"Sounds good to me. There's a new club that's just opened last week that's supposed to be fun. We could start there, and if it's a dud, we can always go to Club Negril. You can't go wrong with reggae and rock-and-roll favorites."

"Okay, it's a plan. I want to go change, so can you come by in say an hour or so? I wouldn't mind a bite of food first. Anyway, it's still early, and I don't think the clubs open until 2100, do they?"

"Yeah, I think you're right. I never go that early, so I don't really know when they do open. So I'll be by as soon as I'm ready. See you soon, love," and after a passionate kiss, she left.

"Anybody home?" Markus asked as he knocked on the slightly ajar door to Natalia's house. He had shown up a few minutes early

"Up here, darling. Make us a drink, and I'll be right down."

"Wow, you get better looking every day," he said as she came down the stairs into her living room wearing a tight red dress with a black silk belt tied loosely around her waist. He kissed her. "Here's your drink. Cheers," and they clinked glasses. They finished their cocktails and left for a night on the town.

CHAPTER SIX

September 7, 2038

Markus left his dwelling with plenty of time to take over his shift at 0800 hours. He gave Natalia a kiss on her cheek before leaving, but she never woke up. He entered the control room ten minutes early and introduced himself to the new person on the desk.

"Hello, I'm Markus, and your name is?"

"Elawa, nice to meet you." They exchanged a few pleasantries and Elawa briefed Markus on the overnight developments. Colin was still moving on a generally westward path, but had slowed down to about five kilometers an hour. Nisha on the other hand was picking up forward momentum and was moving on a northerly path that was making Calcutta the most likely target for landfall.

There were two forest fires burning in the world, one in northern California and the other in China near the Mongolian border. Both were predicted to be under control later today with any help from nature. The forecast was for lower temperatures and the possibility of rain in both areas. There were some minor earthquakes around the globe, but none over 3.2 and no damage. That was about all that happened overnight.

Elawa left after the update, and Markus drew up the morning report, as he knew Kristen would be by shortly for her briefing. He ran some probabilities on both the tropical storms. Colin's was still pretty much the same as yesterday's forecast, but Nisha was now picking up speed, and the numbers were indicating a better than fifty percent chance of landfall near Calcutta. He issued the appropriate warnings for that area. It would still be at least two and maybe three days before the storm would hit on its current path and velocity. Just as he finished, Kristen came into the room.

"Good morning, Markus. How are the storms doing?"

"Good morning. Here's the report. Nisha is looking like Calcutta is its target. Computer projections show a slightly better than 65 percent chance now that it will come ashore within fifty kilometers of the city. I sent out the information to the Indian government and posted a level-three alert for an area five hundred kilometers to either side of Calcutta. Calcutta and one hundred kilometers on either side of it was given a level-four alert. They should begin immediate evacuations of low-lying coastal areas and floodplains along rivers and streams, and start moving livestock to safer zones. As far as the Squad goes, we have three units on hand or in route already to assist in the evacuation. Shelters are being set up away from the coast to house the people who want to leave. I think we might want to consider a level-five alert soon, but I wanted to talk to you about it first."

"What's your reasoning on that?"

"Well, the winds in the center of the cyclone are sustaining a velocity of 260 kilometers an hour, with gusts reaching towards three hundred kilometers per hour. Rainfall could easily hit a half to a full meter or more inland, especially with this cold front we're watching moving down from central Russia. If Nisha runs into that front the rains could be even more intense, although the winds would likely abate somewhat. Flooding will still be near or exceed record levels. I know level five is extreme this early, but millions of people in the floodplain need to get out as soon as possible."

"I see," she responded. "You could be right, but let's wait an hour or so before we send out the alert. Meanwhile, we can do a few more computer models on the situation. If we confirm a high probability for these scenarios, we can have an alert out before most people are up and about anyway. It would be just getting dark there right now, so we could send the level-five alert to the Indian government when we have confirmed the probability, and they could start getting their own people in place. Also, then we could have more Squad units ready to help with evacuations. This might be the best plan of action. We don't want to start a stampede in the middle of the night. Sound good?"

"I agree, that makes the most sense for now. Should I send an alert to more Squad units now or wait?"

"Let's do it now. We might as well have as many of our people on hand as possible. We better start bringing in some reserves too, but let's keep it quiet. Again we don't want to start a panic. I'm sure you have studied what happened in Bangladesh about twenty years ago when a panic evacuation got under way in the middle of the night. Over six thousand people were killed, more than from the storm that they were fleeing. Of course, if they hadn't left, there would most likely have been more casualties, but still it was a terrible waste of lives. That's why we want to do this in such a way as to avoid that from occurring again.

"Also put out a call for medical teams to start toward the region. We'll need food and water distribution stations set up now, so as soon as the storm has passed, those teams can move into the region, but have them stay out of the storm area until we give the all-clear to move. We could use our staging area in southern Thailand for some of the food and water teams. That's only a few hours away by air and appears to be out of the storm track, at least for now. We should also set up some staging areas north and west of Calcutta."

The Squad had set up the use of staging areas through the world for disaster assistance. Most of the time these were abandoned military airfields, as was the case with the one in Thailand. The Squad needed these areas for the movement of the large amounts of food, water, medical supplies, temporary shelters, rebuilding equipment, and so forth that had to come in after any natural disaster. The staging area in Thailand was already fully restocked to deal with the tidal wave from the Philippine earthquake a few days earlier. The rebuilding of the eastern coast of the Indochina coastline was already well under away. Teams had been on the scene in just a few hours after the wave had hit.

"The calls are all out, and we're receiving acknowledgements now." Markus had kept busy while Kristen was talking. "Here's the last one now from the medical teams. They'll be in choppers just out of the storm ready to move in as soon as it's safe enough."

"Good work, Markus. I'll be back in less than an hour. I need to finish my rounds; meanwhile, can you run some more computer projections on Nisha's possible rainfall amounts and run that through a flood forecast? We'll review it when I return. See you shortly."

"No problem, it'll be ready." As soon as she had left the room Markus was quickly running program after program to get the best-case to the worst-case scenario for the Indian Ocean storm. It appeared that even the best case was pretty grim for now. The cold front was still moving southward, and when the computer put those two events together, it showed massive rainfall with widespread flooding in the region. The flooding could reach thousand-year flood levels. That suggested that an even larger area might need to be evacuated.

Markus finished with enough time left to check the rest of the world for problems that September morning. The two fires that they had been watching were just about under control, and not much else was happening, save the two giant cyclones churning away in warm tropical waters and bearing down on land.

Kristen entered the room and he pointed to a screen without saying a word. She studied it for a moment before she began talking. "Has there been any variation in the storm's tracking over the last few hours?" she finally asked Markus.

"None to speak of really. Less than a degree one-way or the other. It essentially has been on a straight path since it came out of that tight little circle it was making. Forward velocity has remained mostly constant also. It should come ashore in about sixty to seventy-two hours at its present speed and course. It will remain a flooding threat for at least that long after it comes on land and maybe even longer."

"Contact the Indian government, and tell them where the storm is now headed," she said in a solemn voice. She knew that the implications of such an action were staggering. It amounted to a mandatory evacuation of the entire area. Of course, some individuals wouldn't leave, and they most likely would be killed in a storm of this intensity. But failure to put out such a call could result in the loss of millions of people and even more animals.

"It's done," Markus stated as soon as he finished putting out the proper code, which always preceded a level-five alert. This way the receiving parties would know that this was not a mistake and to take immediate precautionary action.

"Good, now how's Colin doing?"

"About the same. Still on that westward track and still at about—" He stopped suddenly scanning another computer screen to his immediate left. "Earthquake, looks like central China from these first reports. It's still in progress, so we don't know strong it is yet." Kristen moved around Markus to watch the screen herself as the quake continued. The monitor they were watching was divided into two parts. The top half was a seismograph, while the bottom half was giving out data on location of the quake; population concentrations in the area; any dams or other structures that, if damaged, could result in further danger to lives; the probability of tidal waves; fault structures in the area; the last quake in the region and its magnitude; and so forth. In just moments they would have a great deal of information on the earthquake. It appeared that the quake was beginning to subside.

"Five point two so far," Markus stated. "Not real powerful, but as you can see, there are a lot of earthen dams in the region, which may have been damaged. It's still night there, so it will be hard to check every dam until light. I'll put out a standard call to Squad members in the region and have some engineers there by morning to check on those dams. Let's see what the local authorities are saying."

Markus called the local officials to get their firsthand reports of the quake. When he reached them, he put the call on speakerphone and listened to the woman reporting, "We have minimal damage, mostly broken windows. A few stores had their wares knocked off the shelves, and power is out in some of the outlying areas. Otherwise no serious damage has occurred, and only three reports of injuries. Two people were hurt from broken glass and a third from falling off a ladder while the quake was under way."

"Well, that doesn't sound too bad," Kristen answered back.

"Could have been worse, that's for sure," the woman responded.

"How about the dam?" Kristen asked.

"Many of the dams are quite old and are definitely in danger of breaking," the woman responded. Markus let them know that a team of engineers was already on the way to inspect all the dams in the quake region and would give them a report as soon as they were finished. Also Markus told them that if any dams needed repairs, the engineers would assist them in those repairs.

"Thanks for the help, we really appreciate it," the local responded and then said good-bye.

"Let's hope those dams hold until we can at least check them out. If one were to break overnight … there are a lot of people in some of those valleys downstream," Kristen stated mostly to herself.

Markus answered her concern. "We could put some aircraft with infrared scanning over the site until the engineers have an opportunity to check them out on the ground."

"That's a good idea. Make it so." The infrared on the planes could detect if cooler water was seeping into the dams, so it was very useful in spotting possible damage.

"I'll get right on it," he said, and in just a few moments the order was out to the appropriate personnel.

"At least that way we can see if any are about to break right now. Where's the closest aircraft with that sort of equipment on board?"

"It's coming from Korea, and luck was with us as it was already in the air doing some research over eastern China. They simply altered their course. They should be there within about two hours from their flight projections."

"That's great. Let me know when they reach the site. So where did we leave off with Colin?" Kristen asked, changing the subject now that everything was under control with the quake.

"Let's see … yes, here it is. The storm is still on its westward path and at about the same velocity. It's looking more and more like the islands from Dominica north and westward to Cuba and Jamaica will at the very least get some major winds, rain, and high surf. The track right now would take the storm just north of all of those islands and right into the Bahamas and then into southern Florida.

"But because the storm is so big, it's going to be felt in a lot of areas no matter where the eye of the storm goes. There are hurricane force-winds out in all directions from the center for at least a hundred kilometers. Tidal surges could reach five meters or more. We have watches out in all those areas, and they're just waiting to hear from us on when to start any major evacuations. We're going to be spread pretty thin on help if both of these storms hit within a few days of each other, and it's beginning to appear as they might. Do you want some more reserves called up or at least put on standby?"

"Let's do both. Put out a call for ten reserve units to start toward the areas, and get another ten on standby. Five units on each status are to go to each storm area. The Asian area will need help from outside their region as they're pretty well at full staff right now with the quakes and the tidal wave damage to contend with. Africa, Europe, or Australia would be the best places to draw from for Nisha, and North and South America for Colin."

"Okay, the calls are on the line. We should be hearing back in just a few minutes from the time they receive them. In fact, here's the first one already. It's coming from the Carolinas, and they're contacting their members now for immediate movement to southern Florida. Now more are coming on line. When do you want to put out any higher level alerts to areas in Colin's path?"

"Probably later this morning, as soon as it's within seventy-two hours of the first land areas. Continue with shipping advisories for the entire region and small craft warnings for all areas that might have at least gale-force winds on the storm's present path. Other than that, let's wait until around noon to issue any more watches or warnings unless the storm drastically changes its present speed and direction.

"Keep me posted, I'll be back in an hour or so. I want to be here when the plane reaches the quake area. And patch the plane's scanners through here to us so we can monitor live with them. I don't know what experience they have in quake damage, but our computers can compare the dams' internal structural integrity now to what it was prior to the quake as we have almost every dam in the world on file for this very reason. Any questions?"

"None for now. I'll call you if anything changes. See you soon." Markus made the necessary arrangements with the Korean plane headed toward the quake zone for a live feed to the Squad's headquarters. He further put online a report on all the earthen dams in the affected region so that, when the live feed came through, instant comparisons could be made. He then did his hourly scan of the entire planet, and just as he finished, the Korean plane announced that they were only about thirty minutes out of the quake zone now and would be starting their scanners in five minutes to monitor even outlying dams in the unlikely event that they were damaged too.

"Director?" Markus asked after paging for Kristen.

"Yes?"

"The plane will be over the quake area in about twenty-five minutes. They've started their scanner already so we can monitor all the dams even in the outlying zones."

"I'm on my way. Have you hooked up our—"

"It's already done," he stated, anticipating her question.

"Great, I'll be there in three minutes." Markus had already started the comparisons, and by the time Kristen reached the control room, the first dams were already fully analyzed.

"No damage showing up yet," he told her as she came in, "but we are still about twenty minutes out from the epicenter zone." They both watched as the information from the scanners on board the plane was fed into their computers and the structural integrity of each dam was compared to the report they had on file. Nothing unusual on the first dams was noted. As the aircraft approached the epicenter of the quake, the story changed dramatically. Within ten kilometers of the center of the earthquake, almost every earthen dam was showing some degree of structural damage. Four were experiencing stress significant enough to warrant immediate evacuation of all downstream populations, and the alert was sent to the local authorities. The information they gathered was relayed to the engineers arriving at the quake scene, and closer inspections would commence soon.

When everything was transferred, they returned their attention to the hurricane in the Atlantic and the typhoon in the Indian Ocean. Natalia walked into the control room just as they were looking at Nisha on the satellite picture.

"Good morning. How are the storms doing today?" she asked as she approached them.

"Here's the latest satellite image coming through right now," Kristen replied. "As you can see, it's still a large storm, though not quite as big as Colin, but its winds are just as strong. And the population center it's on course for as of now is far greater than anywhere Colin might hit. And good morning to you too."

"Hi, Markus."

"Hi," he responded as they passed secret glances at each other. "Here's what else happened overnight," he continued, updating her to the events of the quake and so forth. There had been a few small aftershocks in China, and they were closely watching the dams that had shown severe damage. Sometimes it didn't take much more to break one, and the evacuations were not yet completed.

"I'll feel much better when they get all the people out of the flood zone," Kristen stated as they watched the computer screen displaying the quake data. "One of those dams could burst at any moment, especially with these aftershocks hitting."

"They seem to be holding for now," Natalia declared after she compared the latest information coming from the Korean plane that was still scanning the quake zone.

"That's great; let's hope it continues."

"Director, it's almost noon. What about upgrading the Atlantic alerts?" Markus asked, shifting the attention back to the storms.

"Let's take a look at the latest data. Where's the most current satellite image?" she asked him.

"Right here."

"Oiay, still the same course, same speed, but slightly faster wind velocities. I think we better put out the next level alert with special attention to the northern shores of all islands in the area, all of the Bahamas, and most of Florida, especially southern Florida. Would you also pull up all the storm paths for this area for the last fifty years and transfer them to my office, please?"

"It'll be there in a minute," he replied.

"Thanks, I'll be in my office for next few hours, so buzz me there if anything changes. See you two later. Bye."

"Bye," they said in unison as she left the room with her little calico cat right at her heels. Silverton seemed always to be with the director anymore, and if she wasn't, she was with Susan in her new office.

"It's been quite a morning in here, what with the quake and these two storms," Markus told Natalia when they were alone. He finished updating her on all the situations they were monitoring and then officially logged out as she logged in for her shift.

"What do you want to do tonight?" Markus asked her as he was about to leave and go get some sleep. He had sayed up late and of course had been up early to start his watch shift.

"We could go up to the national park and listen to the elk bugle. It is that time of year. Plus the leaves are starting to really turn to their fall colors. Susan went last night and said it was great. I stopped by her office on my way here this morning, and she told me all about it. It's only about an hour and a half drive from here. She suggested taking a picnic and drew me a map of the best places to watch and listen. How does that sound?"

"Great. I'll get a picnic together. Any requests?"

"No, surprise me," she said and kissed him on the cheek. "Now you better go and let me get to work. As soon as I'm off duty, I'll go to my place because I want to change before we go. It's likely to be on the cool side in the mountains, so dress warm. I'll meet you at my place right around then, okay?"

"I'll be there. Bye, sweetheart."

"Bye, darling." He left, and she started her hourly scan of the planet. Nothing had changed since the last scan, so she started further computer projections on the two storms. The rest of her shift was, for the most part, quite routine. She spoke with Kristen several times, updating her on the movement of the hurricanes, which were both still on their same courses, and the latest from China. The evacuations of the areas downstream from the damaged dams were now complete—and just in time: one dam broke shortly after the last person was out of the region. No loss of life was reported, although there was quite a bit of physical damage to buildings, bridges, and other infrastructure. She had sent out calls for help in the rebuilding process, which would soon be under way. The other weakened dams were starting to get repaired as soon as equipment and personnel arrived on site. All in all, the situation was well under control.

Her shift ended and she headed straight home to meet Markus.

"Come on in," she called down to him as he arrived only minutes after she did. "I'm almost ready, I'll be down in a few."

"Okay. Do you have some extra blankets? I didn't seem to have any."

"Yeah, in the hall closet, there's my official picnic blanket. It's red plaid. You'll see it right on the top shelf."

"Found it," he said just as she came downstairs wearing blue jeans and a sweater. "You look great," he said as she slid down the banister right into his arms. They kissed passionately and then headed out the door.

"We want to get there as soon as we can," Natalia said, taking his hand and grabbing the plaid blanket and an extra one.

"The picnic is already loaded, and I brought some binoculars too. With the moonlight, even though there's not going to be much tonight as we're pretty close to a new moon, you can see quite well with them even at night."

"Then we're off," she replied, and they left for the mountains. They drove north, watching a beautiful sunset on their way to the national park. Venus was visible in the western sky while the reddish-orange glow began to fade into night. There wasn't a cloud in the sky. The twilight beauty was breathtaking as they journeyed toward the elk.

"I love you," she said, as she snuggled up next to him in their vehicle. "I've never felt so good with anyone before. It's great, isn't it?"

"It does feel great. I've never been in love like this before myself. I know that we're still pretty young, but wow! You know what I mean?"

"Wow is right," and she kissed him. They didn't say much for a while, just enjoying the stunning sunset and each other. At last they reached the park and drove to one of the sites that Susan had told them about.

"Here it is now. I think we can park right over there and walk up to the top of that small crest," Natalia stated as she pointed to a group of about eight trees atop a ridge overlooking the vale below them. After a quick hike to the crest, they could see some of the magnificent animals starting into the valley from the other side out of the tree-covered mountainside. The night was cool, but not cold. They were dressed warmly, and the cool breeze on their faces was refreshing.

They set up their picnic area and began to feast on a variety of culinary delights that Markus had put together. There was Brie cheese and delicious little crackers to go with it. There was a mixture of vegetables and a dip he had made himself with sour cream, dill, and cilantro. He had also picked up some sushi from one of the many sushi bars in town. For dessert he had a collection of small berries and fruits including blackberries, blueberries, bananas, kiwi, and mandarin oranges. They ate their fill and then sipped on cognac while the bull elks began to put on their nightly spectacular event.

"Look over there," Markus said handing the binoculars to Natalia. "See that huge male? He must have the largest set of antlers I've ever seen. See him?"

"No, I don't see him ... oh, wait, there he is. Wow! Those are some antlers he has. These binoculars work really well," she said, as she peered through them at the animals below. The night air was beginning to get cooler as the breeze shifted to the west, from down off the high peaks. They cuddled under the extra blankets she had brought and soon were in a long erotic kiss. They spent the night on the top of that crest, leaving just in time to get Markus to his shift at headquarters.

CHAPTER SEVEN

September 8, 2038

"Good morning, Markus," Elawa said as he entered the control room.

"And good morning to you. How are you today?"

"I'm great today, but I wish I could say the same for these two storms of ours. Here's the latest. Nisha slowed somewhat overnight our time, but in the last few hours has resumed the velocity it held all yesterday. Winds have increased slightly, and the tidal surges are already starting to show up along the eastern Indian coast. Madras has reported swells to three meters already. That's starting to get up there for the storm still being that far out to sea.

"The other news overnight was that there were two more aftershocks in that China earthquake, which broke another of the earthen dams. Again there were no lives lost as far as we know now, as the area had been evacuated, but more of the other dams are now showing stress, and more evacuations had to be ordered. We sent in an extra team of engineers to help assess the other dams in the area. The Atlantic storm remained basically unchanged overnight our time, and further warnings and watches will be needed later today if it continues on its present tracking.

"Oh, there were a couple of forest fires started by lightning in the past twelve hours also. Both were in the Southern Hemisphere, with one in South Africa and the other in Australia. They are both out of control as of right now, but we have crews on the scenes, and they hope to have a handle on them later today. Let me see if there was anything else. Nope. That's it about it for the overnight report."

"Well it certainly was an active night. Any special directives in operation?"

"No, just the usual ones for a situation like we're in right now. The director is to be immediately notified of any major change in any of the events that we're monitoring. Other than that, nothing out of the ordinary. Do anything special last night?" she asked the Swede, changing the subject as she started getting ready to leave.

"As a matter of fact, I did. Natalia and I went to the national park just northwest of here and listened to the elk bugle. Have you ever done that?"

"No I haven't, but I've heard about it. Are you on tomorrow?" she continued.

"Yes I am. And you?"

"I am too, so I'll see you then. Bye."

"Bye. Natalia and I are going again in a few days to see the elk, and you'd be welcome to join us if you would like."

"That sounds great. I would love to go. Thanks for the offer, just let me know where and when to be ready. I work the early morning shift most of the time; will that be a problem? But I do get two nights a week off, if that would be easier."

"No, we can have you back here in plenty of time, although let me know your schedule for next week, and I'll see if there's a day that you're off that would fit into everyone's plans."

"Okay, and again thanks. It sounds very exciting. See you tomorrow."

"Bye." After she left, Markus did his usual check of the events going on in the world and prepared his report for Kristen when she made her morning rounds. The fires that had started overnight their time were still out of control, and the winds around the African fire were increasing to dangerous levels, so he sent out the next level of alert. This had the immediate consequence of sending two more firefighter units to the sector. It would only take about an hour for the additional units to reach the fire zone as they were based fairly close.

Just as he finished with his projections for all the world's ongoing physical disturbances in progress, the director entered the control room. "Good morning Markus. How are you today?"

"I'm fine, thanks, and you?"

"I'm all right, just a little worried about these two storms brewing out there. What are they up to today?"

"Well, Nisha had slowed somewhat overnight our time, but in the last few hours has resumed its previous forward velocity," he told her; as well as the reports on the fires, earthquake aftermath, and tsunami follow-up information available. They studied the projections on both weather systems and came to the conclusion that the next level of warnings needed to be issued immediately. Markus quickly had the necessary information on its way to the endangered territories. For the Calcutta region, which was already at a level-five alert, this put full evacuation of all coastal areas into immediate operation. All of the eastern side of India, the entire nation of Bangladesh, and western Burma would be in the typhoon's path. Nisha was now nearly one thousand kilometers in diameter, with gale-force winds out approximately three hundred kilometers from the eye. Hurricane-force winds were reaching out as far as 150 kilometers.

This area of the world was historically a target for typhoons. There had been great loss of lives over the centuries in this part of the globe, and even though the Global Disaster Squad was heavily involved in the emergency preparations for these storms, there were just so many people that it would be difficult or impossible to save everyone. They also put out shipping advisories in the entire Bay of Bengal. Flood watches were likewise posted into all of Bangladesh northward through India into Nepal and China.

"I'll be in my office for most of this morning if anything changes. Meanwhile, would you please issue alerts for all reserve units in Asia and North America to be at standby readiness until this is over? Medical units need to start preparations for departure to the both zones within twenty-four hours. There could be some need for medical attention even before the storms make landfall. We just need to be ready. Any questions?"

"No, but if I do come up with some, I'll call you. Anything else?"

"Not at the moment. I'll be back before your shift is over. See you in a little bit," she said and left the control room. Markus went straight to work on her directives.

It was very probable that the Asian storm would make landfall first. Colin, in the Atlantic, would begin affecting the island nations within about twenty-four hours after Nisha went ashore somewhere near Calcutta. It would then be a day or maybe two after that before Colin struck the mainland of North America, with Florida still the most likely target on the storm's present path. After it crossed the peninsula of Florida, all the land areas surrounding the entire Gulf of Mexico would become a possible target. Advisories for that region wouldn't be issued until the storm was actually approaching Florida. Then watches and warnings would be put out onto the Web immediately. There was still the chance that the storm would veer north and never enter the Gulf of Mexico. So until it was fairly certain that this wouldn't happen, no further alerts would be issued.

The overall status of the storm situation was one of apprehension. The range of possibilities was still quite large. Best case would have the storms move into colder waters and dissipate over the oceans. The worst case could be really bad with great loss of life, major forest and natural habitats damaged, plus destruction of homes, businesses, and infrastructure.

Some Atlantic hurricanes had caused extreme loss of life and property damage over the last fifty years. For example, in 1992 a storm named Andrew went through the Homestead area south of Miami and caused more damage that any other natural phenomenon to that time in history. It later crossed the Gulf of Mexico and went ashore again in the Louisiana area. There the storm caused a great deal of flooding and misery. Then there was Katrina in 2005, still a textbook study in how not to handle a natural disaster. There was even a course at the Squad academy studying the way that storm was mishandled. Other storms since then had also resulted in great physical destruction in parts of North America.

Markus continued his prognostication of the situation, running every possible computer model he could conceive for the two storms. Still the highest probabilities put Nisha ashore at most thirty to fifty kilometers west of Calcutta. In terms of a storm of this magnitude, missing Calcutta proper would cause as much damage as a direct hit.

He was about to start another series of computer runs when Natalia dropped by the control room. He looked up with a smile. "Good morning, sweetheart. How are you?"

"I'm great. Thanks for a really great time last night. I enjoyed it thoroughly. I was listening to the news and saw the latest on the Asian storm. I just couldn't wait until my shift to see the details, so I came on over."

He explained Nisha's current position and the alerts that the Squad had issued just a little while ago and brought her up to date on all of the world's plights for that morning. As they were looking over some new data coming off the computer, the seismograph began flashing to indicate an earthquake in progress. This was a new one.

"It appears to be in South America. It looks like southern Colombia, approximately five hundred kilometers southeast of Bogotá, in the mountains. I'll get the first-level alert on the wire," Markus said as he quickly executed all the appropriate procedures for an earthquake, which included notifying local authorities of the exact location of the quake, any extraneous situations such as earthen dams in the quake area, tsunami probabilities, and alerting local Squad members with the same information. He then called Kristen's office, and in moments she was on her way to the control room.

"How big?" she asked as soon as she entered the room.

"It's still going on, but the largest reading so far is six point nine. It's a strong one. Its epicenter is right about here," he said, indicating a map point on the computer screen. "The first-level alerts are already on their way, and I put out a next-level call to medical teams in the region. There is quite a population scattered through that part of Colombia, although there are no large cities or population concentrations. As in the China quake, there are concerns about earthen dams breaking."

"Better get some engineers on the way, and see if you can find a plane with the same equipment as that Korean aircraft and start it toward the area."

"I'm on it."

"Good morning Natalia," Kristen said after the initial responses over this new quake were under control.

"Good morning. I just wanted to see what was happening with the two storms, and then this happened. There sure have been a lot of quakes lately, or does it just seem that way?"

"I was wondering that myself, so I checked the frequencies over the last one hundred years. I found that there have been thirty eight years with more total earthquakes, but only eight with more severe quakes. And on top of that, the really violent quakes have all occurred in a very short time frame, plus the tsunami this year with the Philippine quake has made this quite a year for seismic activity. So to answer your question, yes, this has been a fairly active year for quakes, but not the most. Anything more in the year could, of course, change that. But let's hope that's not the case."

"That's for sure."

"It seems to be receding," Markus reported, as they all gathered around the computer screen to watch a live report coming into the control room from the first Squad members on scene.

"This is Squad member Paulo Niman; do you copy me?"

"We do. This is Squad headquarters. Wwhat do you know?" Markus replied.

"We have reports of numerous fires out of control in several small towns in the quake region," Paulo replied. "This is bad news, thanks to many months of drought in this part of South America. The fires are likely to spread through the forests is high."

Markus called in firefighter units the moment he heard the news from the field and gave them the coordinates of the blazes.

Paulo continued with his report. "We are getting numerous smaller aftershocks still hitting the area, I can feel them right where I am. I just received a report that a dam has broken, and it is feared there will be casualties downstream. We need evacuation orders now in those areas. I am sending you the coordinates right now. Also, can you get us a plane here to scan all the dams in the quake area to check for damage?"

"We're on it now," Markus responded, and after sending the evacuation orders, found an aircraft with infrared detectors on board and had it in route in just minutes.

"It will be there in less than an hour, he added.

"That was fortunate," Natalia said, after hearing that the plane was so close. The plane was in Peru right across the border.

"Get some medical help on the way to the flood areas," Kristen said.

"Teams are on their way," Markus declared after he sent out the call for help.

"Headquarters, this is Paulo again. I am hearing now of from seven to twelve dams that are leaking or possibly about to break," he reported. "I think we should evacuate all downstream regions right away as a precautionary move. The damage seems so widespread that waiting might be disastrous."

"I think you're right," Kristen said. "We'll send out the evacuation alert right away. That plane should just about be there."

"Yes, it's entering the airspace of the quake region now. We're on automatic feed from the plane, and here comes the first data." They started studying the information that was coming over the computer. The computer was again comparing the plane's scanning with the data they had on file about every dam in the region. Quickly it showed that many of the dams were damaged, and some were in severe danger of collapsing at any moment. Markus sent out this information as soon as it was available on each dam in the quake zone. On-site engineers were confirming their data as they entered the area. This had the potential to be even worse if there were any severe aftershocks. The evacuations needed to be completed before any more tremors hit.

Paulo Niman, from Costa Rica, was the senior Squad member present. He had been a member of the Squad for almost twelve years and was in charge of southern Mexico, all of Central America, and northern South America. It turned out that Paulo was near the quake site and had moved near the epicenter when he had started making his reports.

Paulo continued with his report. "There are as many as one hundred dams in the region that could break with any more aftershocks. Numerous bridges are also showing signs of stress. Some bridges have already collapsed, while others will not be able to withstand either another aftershock or a great deal of traffic on them. And unfortunately they are also the only way out of some of the evacuation areas. Whoa, we're getting another aftershock," Paulo exclaimed.

"How big?" Kristen asked Markus after hearing Paulo.

"Five point nine."

"Paulo, you still with us?"

"Yes, I'm here. That was quite a shake. Some buildings collapsed that I can see from here, and there's a crevice in the road right in front of our communication center. It looks like a lot of the structures were weakened, and this was all they could take. I'm getting radio reports now from our other Squad members scattered throughout the quake area. Same deal, buildings down, some fires have started, and at least three dams have broken, according to these first reports. Several bridges down, it looks like it just went from bad to worse, Director. Can you send more help?"

"What do you need?"

"Medical, mostly. I'm getting descriptions of collapsed buildings with people trapped in them. There was a bus on one of the bridges when it went down, and first accounts from the scene show maybe as many as fifty to one hundred dead or severely injured in that accident. Also one of the three dams that we know collapsed has sent a wall of water through a small town of three thousand people. We have no word from the town itself yet; these are helicopter reports from our scouting teams. It doesn't look good there. I'd estimate that we've lost up to five thousand people so far, and that number could go higher as we start digging out these buildings. We also could use more personnel to assist in the digging out and appropriate equipment to expedite the job. We're going to need water purification systems too. So much of the drinking water has been or will be contaminated from the breaking of the dams that a lot of the water won't be safe to drink. That should do it for now. Anything else we need to know from your perspective?"

"Not at the moment, we'll keep you up to date on anything new that comes up. We'll talk to you this afternoon for a routine briefing if we haven't talked before then. Let us know if there's anything else we can do. Good luck, Paulo."

"Thanks. Talk to you all later."

"What a morning. I'm going back to my office and finish a report I need to have out before noon. Meanwhile keep me posted as always. See you two later," Kristen said and immediately left the control room.

"This has been quite a morning," Markus echoed once they were alone. "First these two storms, the fires in Africa and Australia, then the earthquake and look here—more storms."

"Tornadoes in northern France," Natalia said. "That's sort of rare for this time of year. Seven dead, hundreds injured, fifty-plus homes destroyed and many businesses damaged also. The local medical teams have the situation under control, and the Squad is already on the scene with repair estimates under way."

"How's Nisha doing?" Markus asked as he turned to the computer screen monitoring the Indian Ocean typhoon.

Natalia scanned the incoming data on the cyclone. "It's still on the same track, and velocity has remained unchanged. Gale force winds have extended a little farther out from the eye of the storm, but otherwise no significant difference from last hour. You might as well go rest, Markus. I'm ready to take over my watch now, and there's no reason that you need to stay. What do you want to do tonight?"

"Isn't tonight Susan's get-together?"

"Oh, I think you're right. Want to come by my place first, and then we'll go on over? She lives pretty close to me anyway; in fact, we can walk on over."

"Sounds great. I think I will leave now and get some sleep. Take care, sweetheart, see you later." Markus signed out and gave Natalia a kiss on the nape of her neck as he left. She got goose bumps down her spine and shivered in response.

"Come over as soon as you can," she said as he left.

"I will." Natalia went to work with her normal duties and began to do more computer projections on the two storms. There was no substantial change in the predictions and she noted this in her log. The rest of afternoon was fairly uneventful, which in this business was good news. There were a couple more tornadoes in that band of storms crossing northern Europe, but only minor damage had resulted and no reports of any injuries. There was that forest fire in Mongolia, now just about totally under control; a new little fire in Canada was quickly surrounded, although it would take several more days to completely extinguish. The fire in southern Africa was getting under control thanks to some rain moving into the region.

The rebuilding from all of the last two months natural disasters was on schedule with the exception of some of the tsunami damage in Southeast Asia. Heavy rains in that part of the world had slowed the movement of goods on top of the fact that the entire coastal highway had been destroyed, including most of the bridges. The rains were predicted to end in another week or so, and the Squad hoped to get back on a revised schedule soon.

There had been a severe dust storm in southern Africa ahead of the front that brought the rain, which helped control the fire there, but there were no reports of damage or injuries. An interruption of radio communication in the area had lasted only a few hours. The sunsets, though, were absolutely spectacular, according to on-site reports.

The fires triggered by the earthquake in Colombia were still burning, but again it appeared as though Mother Nature might be able to lend a helping hand. After months of drought, there were signs of a large rainstorm developing off the coast in a new El Niño, which had just recently formed off the South American coast. Throughout the rest of the world it was a pretty quiet fall day in the Northern Hemisphere and a beautiful spring day in the Southern Hemisphere.

"How's it going?" Kristen asked as she entered the control room.

"Hi. I was just sending you an update on the world's hot spots," Natalia said as she opened a file on her screen. "Here's the summary of what's happening. Looks like Nisha is strengthening somewhat. Gale force winds are extending further out from the eye of the storm, and I've sent that information out to the areas under typhoon watch. Colin is remaining pretty much the same intensity with only minor fluctuations in wind velocity. I already sent the reports on the fires burning, the windstorm in Africa, the tornadoes in Europe, and a follow-up on the Colombian and Chinese earthquakes. There have been no further aftershocks of any size so far behind either of them. Relief efforts are in full swing, and the additional medical help is on the scene already in Colombia. One more dam did break, but the evacuation downstream was complete before it gave way.

"The downside is that there are reports of a lot of casualties coming in from the initial dam failures. Many villages were completely swept away, and it appears that there were few if any survivors. We may never find all the missing; they very likely are buried under meters of mud and debris. Also the census records in some of these villages were not very detailed, and what records there were might have swept away in the flooding that followed the dams breaking.”

“Well, we’ll just do the best we can. This happens quite often in the more remote regions of the world. What about Nisha? When is projected landfall?”

“In about forty-eight hours.”

“Who’s the senior Squad member in the region?”

“Let me check … here we go: Helene Jansson, on station in Calcutta. Why look, it’s Markus’s cousin. He told me that there are three members of his family in the Squad, including a cousin and an uncle. His uncle is the chancellor at the academy. I’m sure you know him.”

“Indeed I do. I’m in constant contact with him. We have to attend many of the same functions, meetings, budget conferences, et cetera. In fact, we’re meeting next week on some budget questions. He’s coming here, so you’ll get an opportunity to meet him.”

“That would be great. I’ve met him of course at the academy, but it was very brief, and he must meet so many cadets.”

“You’d be surprised to know how much he remembers about every single cadet who graduates through the academy. He takes great pride in every person who completes the rigorous training at the academy. Do you know Helene?”

“No, I don’t, but I’ve heard Markus talk about her. He says she’s quite the adventurous one. I think he said she’s been in the Squad for about eight years. She’s five or six years older than Markus.”

“Well I wouldn’t doubt that most of the stories are true. I’m glad she’s the one in charge on the site. Can you contact her and patch it through to my office, please? I want to go over some of the details of the evacuation and so forth. I’ll be there in about ten minutes, as soon as I finish my rounds.

“Thanks. I’ll see you tonight I think, right? Aren’t you two going to Susan’s?”

“Yeah, we’ll be there. See you later.”

"Bye," Kristen replied and left the room. Natalia waited ten minutes and then put a call in to Helene and connected her to the director's office as soon as she was on the screen. A couple of minutes later Paul came in to relieve her of duty and she went home to get ready for Susan's party.

CHAPTER EIGHT

"Wow, I've got to be the luckiest guy in the world," Markus said as Natalia came down the stairs. She was dressed in a slender black silk dress with a gold belt and gold earrings dangling halfway to her shoulders. Her hair was tied back in a French braid loosely held with a flowered hair clasp. She looked stunning. Markus walked over to her, took her in a tight loving embrace, and gently kissed her red lips. "I love you so much," he whispered into her ear.

"I love you too," she murmured back to him through the kiss. They finished their cocktails and headed over to Susan's. It was only a couple minutes away on foot, and soon they were knocking on door, which stood ajar.

"Hi, guys, I'm glad you made it. Come on in," Susan said as she greeted them. She introduced them to her friends that they didn't know, and they mingled through the guests. They saw Kristen talking to someone that they had just been introduced to as Susan's visiting friend that she had wanted them to meet.

"Hi, you two, you won't believe this. Remember that I wanted you to meet my sister who was going to be visiting here next week. Well, she's here already and turns out to be a friend of Susan's too. In fact, the two of them planned this little surprise for me, as Susan never told me that Joan would be here. Have you all met?" Kristen asked.

"Yes we just met as we came in," Natalia replied. They continued talking with the director and her sister Joan for a while discussing the current world storm situation and what it was like to be a shuttle pilot. She made about three trips a day and was scheduled to be the first pilot to land at the new spaceport in southern Africa. The transport system stemmed from the United States shuttle program of the last century. Now there were seven and soon to be eight spaceports for the shuttles to fly to and from. This form of travel had greatly reduced the time that it took to travel around the world. Most flights were under two hours. They were, simply put, little excursions into space and then back into the atmosphere above their intended destinations. They had only been available for commercial flights for about twenty-three years and the cost of the flights had kept most people away at first. But now technology had reduced the cost of the flights, and many more passengers now flew the shuttles quite regularly. It was very exciting to take one of the flights. The view from space was breathtaking.

They also discussed Joan's legendary namesake relative, Joan Osborn, who was a famous singer in the late twentieth and early twenty-first century. She was still active in the music industry and had just recently released a duet album with some of the young artists of today. It had reached number one on the Billboard and Super VH1 charts. The new compact disk was being mentioned for at least five Grammy nominations. Both Natalia and Markus were fans of Joan's namesake and hoped to see her in concert someday. Joan told them that when she came through this area in a few months, she would get them tickets to the show, and she would plan to take them to the concert herself and take them backstage after the concert. They thought that was great and thanked her for the offer. They exchanged phone numbers and agreed to keep in touch.

After a bit Natalia and Markus got hungry and went to eat some of the great food that had been put out for the guests. Susan had gone all out for her party. The food was elegantly presented and deliciously prepared. The cuisine was vegetarian in essence, although there were cheeses, egg dishes and shrimp served Cajun style. It was quite tasty. The guests also enjoyed a full bar with several wine choices representing five continents and beer from twenty-three nations. The view from Susan's home faced the Rocky Mountains, and the sun was just beginning its descent over the mountains. It was a clear evening and the sky was a brilliant scarlet tonight with the high dust that was common here this time of year. It was incredibly beautiful, and all the guests had gathered at the large windows to observe the spectacularly colorful event.

"Look," Markus said and pointed to a small crescent moon, which was just becoming visible in the waning sunlight. To the left of the moon was the planet Venus shinning brightly in the September sky. It was a breathtaking view. "I think that's my favorite view of the sky," Markus continued. "The crescent moon and Venus always look so great together."

"Mine too," several of the others guests chimed in unison.

"We're bringing out the desserts now," Susan announced after the slender crescent moon had set below the mountains and Venus had quickly followed while the night spread across the sky marked with the sparkling light of the stars. The guests made their way into the dining room where the table had been transformed into a sweets lover's dream come true. There were almost thirty different treats to choose from, including pies, cakes, truffles, pastries, puddings, cheesecakes, and fruits from around the world, kiwi sherbet, and so forth. Again it was a masterpiece.

Everyone was greatly enjoying the gourmet delight when Director Verba's pager went off. She had it set on vibrate so that only she knew it had gone off. She took out her cell phone and went into a small study to call the control room.

"Global Disaster Squad, Paul speaking. How can I help you?"

"Paul, it's Kristen. What's up?"

"There's been a dramatic change in the Indian Ocean storm. It has increased forward speed and will make landfall sometime late tomorrow if this continues. I've notified the governments of the nations that will be affected and also called Helene to tell her that the evacuations must be completed twenty-four hours sooner than she had originally thought. I thought you would like to know."

"Yes, I do want to know. Thanks, Paul. I'll be over there in about thirty minutes, as soon as I say good-bye and drive over. Bye." She went back into the room with the other guests and was looking for Susan when Markus came over to her.

"Something up?" he asked, sensing a slightly hurried look in her eyes.

"Yes. Maybe you and Natalia had better come over to headquarters as soon as you can. It's Nisha. The storm has increased forward velocity significantly and will make landfall some twenty-four hours ahead of what we originally thought. It might turn this into a major disaster. I don't want to wreak Susan's party so I'm not going to make a big deal out of this right now. After all it may slow down again anyway. I would suggest that you two do the same. If you two came in an hour or so that would be fine. Oh, there's Susan now. See you over there, bye."

"Okay, later," Markus responded. Kristen gave her thanks to the hostess and said good-bye to several of the other guests and of course her sister Joan, whom she invited over to Squad headquarters after the party. After thanking Susan again for having Joan there, she left for the Squad's headquarters.

"Anything change?" Kristen asked as she came into the control room.

"Nothing at the moment. Forward velocity is steady at nearly double what it was earlier, and landfall is expected by tomorrow night if this continues. We've sent word to all affected areas to speed up evacuations. This is an urgent situation. There are still four million people in very dangerous coastal regions."

"Can you get Helene on the screen, please?"

"Sure. In fact, I just talked to her only a few minutes before you came, so she's probably close by. Here she is now ..."

"Yes, this is Helene. Is that you, Kristen?"

"Yes it is. What's your situation? How are you doing?

"I'm doing fine, but this place is a mess. People are starting to get a little anxious. I've tried to keep the evacuations under control, but the situation is deteriorating rapidly. There already have been some major problems. All roads leading out of here are jammed. There are reports of two bridges collapsing under too much weight, and the winds have already picked up. If you could get some choppers in, maybe we could get some of the pressure off the roads. Also some heavy transport planes could get a lot of people out of here in a hurry along with their livestock. Animals are really blocking the rapid movement of vehicles on the highways. We've got every train, bus, car, motorcycle, you name it in the area carrying people out of here, but there's so many. We're concentrating in the low-lying areas first, but I'm afraid that the storm is going to impact far into the countryside.

"Rain predictions indicate severe flooding is likely throughout this entire region from Cuttack, west of here in India, to Chittagong in eastern Bangladesh. Flooding could reach far inland too, as far north as several hundred kilometers, because the flat terrain will hardly weaken the storm for some time. All along the Ganges up to Bhagalpur, India, we have evacuations under way to any higher ground available. The situation is worst, of course, closest to the Bay of Bengal. The flooding higher up the river probably won't start until two or three days from now. It's the tidal surges that we're most worried about now. Still several million people are in those low-lying coastal regions and that's where most of the effort is going on right now."

"We'll get those transport planes in there as fast we can. Where do you want them?"

"I'll send you the coordinates in just a few minutes. We're working on where the highest concentrations of people are right now, and we should have that data to you right about ... now."

"It's coming online, and we're relaying it to the planes, many of which are already in the air. Can we divert more trains into the region?"

"The tracks are pretty crowded already, and we have one broken-down engine on a side line that's holding up a lot of other trains. We're just trying to get it out of the way right now. That's about all we can do. Unfortunately it broke down on a part of the track that is not near any switchouts. We might have to just topple it over to get it out of the way. We're making one more effort to get it running, and if that fails, we're going to get it off the tracks."

"What about more buses?"

"The roads are so full now that we can't get any more vehicles on them. The only real hope we have is to airlift people out of there. Are there any more choppers close?"

"We can get some from southern India in a few hours and a few more from around the area. But there aren't a lot within six hours and any longer away there won't be enough time to use them."

"Well, any that we can get here will be better than nothing."

"That's for sure. We'll do the best we can and get as many to the area as possible."

"Sounds great. Good news, though, that locomotive is now working, and we've got it moving already."

"That is good news. Well, keep us informed. We're going to see what we can muster up from anywhere in the region. I'll call you back in about an hour. If you need anything before that, please feel free to call me. Talk to you later."

"Thanks, talk to you soon. Bye." The screen went blank, and Kristen turned to Paul.

"Paul, would you put out the highest-level emergency alert to get every available aircraft within twenty-fours of the area to start picking up evacuees and getting them out of there? If it can fly, let's move it there. Then alert the air control people to start an in-and-out flight pattern that maximizes the available air space. Also find out the closest places where we can take this number of planes that will be out of the storm's main path. Markus and Natalia are coming here soon to assist us. I think we may have to go to level five here by midnight our time."

Level-five alert for the Global Disaster Squad headquarters meant that all available personnel moved into the control building until the crisis was over. This allowed for more than one person monitoring events at a time and for the others to expedite any needed peripheral actions. It was always necessary to have someone scrutinizing the rest of the globe even while the majority of the Squad was maintaining a high level of vigilance on the current situation. With Nisha so close to landfall and Colin quick to follow, Kristen thought it would be prudent to go to level five within a few hours. All Squad members carried a beeper that kept them in constant touch for just such emergencies.

"Paul, let's go ahead and put out the level-five alert to all headquarters personnel, effective midnight our time."

"Done," he said pushing the control switch, which automatically sent the call to all the members. Squad members had one hour to get to headquarters after a level-five alert went out, but most would be there before that. Within just minutes, the first Squad members started checking into the control room, and the energy level went up as these highly trained individuals prepared for the worst, knowing that they would have been called on a level-five alert only in the most dire circumstances.

"When everyone arrives, Paul, we'll have a briefing in the ready room. I'll be back in less than an hour and that will be time enough for most people to be here. See you shortly." Kristen wanted to go to her place and change clothes out of her party dress and arrange for her two cats to be taken care of should she be stuck for several days at headquarters as she was expecting. She called her friend Sarah, who had a small business in the city and was not a member of the Squad to watch her cats. Grabbing some personal items she hurried back to the control room. By the time she returned, the majority of the others were waiting in the ready room. It was quite a collection of people, experts in every conceivable field of science, logistics, linguistics and anything else needed to carry out the responsibility of the Squad.

She went to the front of the ready room to begin the briefing. "Thank you all for your quick response to the level-five call." She explained what was occurring and assigned each person to a task. They set up a work schedule of twelve hours on and four hours off for the time being with a staggered timetable so that there would always be at least four Squad members on duty at any one time. Some of the members were assigned to the Atlantic storm, and it was decided that evacuation alerts should be increased one level in view of what was happening in Asia. They didn't want to be caught in this situation again. Still others were given the duty of monitoring the rest of the world. This would be quite an intense few days coming up.

"Any questions or comments?" she asked after she was through with her briefing. There were a few, and after she had dealt with them, the group broke into their assigned areas. Markus and Natalia were given the first watch on the Atlantic storm. They went to the auxiliary control panel, which was in the same room as the main control area, but was not used unless they were in a level-five situation. Quickly they put out the higher evacuation alert code to all affected regions.

The storm was still moving at about the same velocity, but it was getting larger, and the barometric pressure was still falling, driving wind speed higher. Hurricane-strength winds now reached out about 150 kilometers from the center of the storm, and gale force winds reached three hundred kilometers from the eye. Tidal surges could be expected to start hitting the outer Leeward Islands within about the next twenty-four hours, and small-craft warnings were issued for the entire region northward to the Carolinas in the United States via the WMO (World Meteorological Organization). There would likely be choppy seas in a very large area, and all small craft were advised to move to safe harbors or out of the area completely.

Elawa was stationed right next to them with the job of monitoring the rest of the world. There was still the Colombian earthquake to watch. Five more dams had collapsed in minor aftershocks since the last main quake. Rescue efforts were being hampered by rain, which had become torrential just in last few hours. The drought was finally broken, which they needed, but not this much rain so fast, and the need for medical help was still critical in some of the flooded zones downstream of the broken dams.

There were also eight fires now burning across the world, mostly in the dry autumn of the Northern Hemisphere. Three were in North America, one in Europe, two in northern Asia, and one in southern Asia. The last was a new one in the Southern Hemisphere, in the African nation of Zambia. At the moment they were all out of control, but two were expected to be under control within twelve hours. The others were still burning unchecked. One in northern Asia had already scorched several thousand square kilometers of forest and was being fanned by fifty-kilometer-per-hour winds. It was definitely the most serious fire at the moment. All these events in Asia were creating an aircraft shortage in central Asia as most planes were being diverted to either fighting the fire or evacuating people out of India and Bangladesh.

"What's happening in Colombia?" Kristen asked Elawa as she went through the room to check the current situation in the world. Elawa brought her up to date on the earthquake and the relief efforts going on to rescue as many people as possible. But the rains were really starting to make the situation desperate. Many roads had been destroyed when the dams broke and sent walls of water down the valleys, and now the rains were keeping water levels high and threatening other roads, bridges, and dams. The good thing about the rains was that they put out the fires that had been started when the earthquake first hit. Aircraft were having a hard time flying in the rain as a lot of the airports were on limited or intermittent power for their radar. On top of that, some airports were now underwater themselves. This, of course, had made landing an aircraft impossible.

"What else can we do?" Elawa asked after bringing Kristen up to date, concern obvious in her voice.

"We need to get some boats into the flooded region by morning. We could airlift them in with helicopters and drop them into place without having the aircraft actually land. Then they could get to the stricken areas by water, since there seems to be so much of it. See what you can do about this, Elawa, and I'll be back to you in a bit. Also see if there are any hydroplanes nearby. We could always use them on the larger water areas such as lakes, rivers, or even some areas of flooding."

"Okay. I bet we can find something somewhere that can do the job. Talk to you soon."

"I'll be back." Kristen moved on to talk with Natalia and Markus. They reviewed the Atlantic storm and concluded that even another level of alert was not out of line. Markus sent the appropriate communication to the WMO after clearing it with Kristen.

Looking around the room, Kristen saw a room full of individuals working as one, focused on their one task: save as many lives as possible.

CHAPTER NINE

The sun was just breaking the eastern horizon outside the Squad's headquarters, splashing its bright cheerful colors over the slightly clouded sky. September in Colorado could be one of the best times of the year there, with its warm days and clear crisp nights. This year was a little warmer than most but was still having cool nights that made sleeping very comfortable.

Unfortunately for everyone at Squad headquarters, no one had slept enough to enjoy the comforts of the night. The day was just transforming into night in central Asia, where it was almost exactly twelve hours later than Colorado. The rain had already started to fall from typhoon Nisha, and the winds, while not very strong yet, had definitely started to blow. The quiet before the storm, which had hung in the air just hours before, was now gone for good. The storm had begun in earnest, and it was only a matter of time until the full fury of it was to bear down on this crowded section of the planet.

Kristen had finished her work sometime after midnight and had then taken a few hours nap. She woke with a start. Realizing that she had been dreaming, she calmed herself down and, with her eyes shut again, thought about the upcoming day. She then remembered that it was her parents' wedding anniversary and jumped up to call them. She computed the time change and realized she had just eight minutes to reach them still on their anniversary. It was eight minutes to midnight in eastern Australia. Quickly she punched their code into her communicator, and while her message streaked into the sky and back down to her parents' home, she hoped they would still be awake. She thought they would be as it was their custom not to go to bed until midnight on special days such as this.

Her mother answered. She chatted with them for fifteen minutes about all the rest of the relatives, her immediate family, her social circle, her friends back in Queensland and her work. They were very proud of their daughter, and the fact that she was the head of the Global Disaster Squad was especially important to them as they had both served in the Squad when they had been younger. It was where they had met and fallen in love. They secretly, or so they thought, hoped the same would happen to their daughter and that she would find a man to fall in love with just as they had done years earlier.

After ending the call, Kristen had a moment of nostalgia thinking about her life in Queensland and how she hoped to visit there again soon. She missed her family and friends. The thought was quickly stored away when a voice broke the morning silence over the loudspeaker in her office.

"Director, are you there?" Paul's voice asked.

"This is Kristen, go ahead."

"I think you need to come to the control room right away. There's been an increase in Nisha's forward velocity in the last half-hour, and it seems to be still increasing."

"I'm on my way." She was in the control room in just minutes and went directly to the monitor watching the Asian storm. The continuous updating of information was flashing across the computer screen, and she asked what the new velocity would mean to the landfall projections.

"Since the storm is already fairly close, it won't change much if the speed doesn't increase much more than it already has," Paul answered, still on duty. He was the first to finish a twelve-hour shift and was still on the job.

"Isn't your watch over by now?" she asked him realizing that he was the one on duty when they went to level five last night.

"Yes, it is, and I'm going to go and get some rest in a moment. It's just that I get so involved I don't want to miss anything."

"I know how you feel, but we're going to need you back here later when this really gets going, and we're going to need you rested. So please, go and get some rest. Believe it or not, we can take care of it while you're gone," she said teasingly.

"I suppose you can," Paul responded with a laugh and left the control room for some much warranted rest after his replacement took over the watch station. Indira was no stranger to the monitoring stations though, having worked them for the first three years of her tour of duty.

"Indira, good to have you on board again," Kristen said, welcoming her back to the control room.

"Thanks, it's good to be back, but I wish under better circumstances. So we've got some big ones bearing down on the world right now."

"I'm afraid so. These could get really out of control if they stay as intense as they are right now," Kristen replied.

"Nisha looks like it's slowing down again," Indira said after running her first diagnostic on current information. "See here, it's not much, maybe a kilometer or two per hour or even three, but it definitely has slowed. That's good news for now, when every extra moment can mean more of the evacuation complete, but obviously this storm is quite erratic when it comes to forward velocity."

"That's for sure. We're racing the clock, and the first of the winds and rains have already started slowing the process down somewhat, although it's not real bad yet. What does the computer say now about landfall?" Kristen asked Indira.

"Let's see here, I remember how to do this—here it is. Looks like about twelve hours left at this current speed. That would put it there around dawn local time or sunset here," she answered Kristen.

"That's still some time. Can you raise Helene for me, please? She's the senior Squad member on the scene and in charge of the evacuations."

"Sure. It will take just a moment. Someone is going to get her now."

Helene was making rounds of the India control center, checking on evacuation progress, current wind velocities, rainfall amounts, and other information when she was told Squad headquarters was asking for her. She went to the communication center to answer.

"Kristen, is that you?"

"Yes it is. How's the evacuation going?"

"We're still moving thousands of people per hour per square kilometer out of all the low-lying regions. We still see people in harm's way and are going as fast as humanly possible. What's the current landfall forecast?"

"Sunrise your time. The storm had increased forward velocity for a while, but now seems to be slowing again. We're watching it very closely and will be sending you continuous updates on the half-hour until midnight your time and then every ten minutes after that until it actually comes onshore. There are eighteen more transports within an hour of your location. These are those enormous new ones that can hold close to a thousand people per plane. Let us know where you want them first. The sooner the better so we can start routing them accordingly."

"Sounds great. What runway specifications do they require?"

"We're sending that data to you now. They employ some new technology that allows them to land on runways quite a bit shorter than you would think. Our comparisons show at least five to eight airports in the coastal zones could accommodate them. Are many people still at the airports, or are they trying to move inland over the highways?"

"Some have started to move by land, but we've still got tens of thousands of evacuees at the airports. So with these new planes we can move close to eighteen thousand people per trip and if we can complete a trip every two hours including loading, unloading, and flight time, we can get another ninety thousand out of there before dawn. We have several airports within half an hour by air of the low-lying areas that will be safe so we can get as many round trips as possible. That leaves only one hour for loading and unloading a thousand people. That will be the hard part, as many of them have never flown in a plane before, and they're frightened, tired, and so forth. But we'll figure out something to keep them moving. Any other news we could use?"

"Nothing else for the moment. How are the trains holding up?"

"They're running virtually nonstop. We fill them to overload capacity, move to higher ground and only stop for five minutes; then they turn around while people are still getting off and return for the next load. At the other end we stop for only five minutes to load, but the people are used to that on an everyday basis so we can fill an entire train in less than five minutes. We're moving thousands of inhabitants per hour. We're pretty much just running the trains up to higher ground and then coming back for more. The round trip is about an hour average for all the trains."

"Are you going to be able to get everyone out before sunrise?"

"It'll be close, but I think we'll have most out for sure. We'll still be moving people even after the storm passes to be ready for the flooding that will follow the passage of the typhoon. Right now we're concentrating on the coastal lowlands and immediate river flooding zones. If we can have those regions cleared before sunrise, we have projected we'll have up to another eight to twelve hours before the rivers will begin to reach the first flood stages.

"Of course, a lot depends on what the storm does after it comes onshore. If it lingers, that creates one set of problems, or if moves quickly inland and rains itself out further upstream, then the rivers may flood more downstream. We'll just have to keep monitoring everything very closely.

"I've got to go for a while. I'll check back in with you in about an hour. Does that sound good with you?"

"Sounds good. Talk to you soon. Out." Helene took a deep breath knowing that the next twenty-four to forty-eight hours would be beyond intense, and returned to giving direction with the latest information she had just received. She needed to have everyone pick up the pace.

"She sure knows what she's doing, doesn't she?" Natalia said after Helene broke off communication with control headquarters.

"Indeed she does," Kristen responded, adding, "She's been with the Squad for quite a while, and her passion for it hasn't diminished one iota in all those years. Her secret is that she loves it. Let's hope she can pull off this evacuation."

"Kristen, look at this," Indira said. "This is just coming in through the computer. A band of waterspouts has formed right off the coast south of Calcutta and appears to be moving north-northeast at about twenty kilometers per hour," She sent the data to local authorities in India and to Helene.

"Just what we didn't need. Better get that right out to—"

"It's already done, Director," Indira said with a twinkle. "I haven't forgotten that much since I worked this room."

"I suppose you haven't. Well, good. What's landfall projection on the waterspouts?"

"Less than an hour. They're about fifteen kilometers offshore right now and if they hold their twenty kilometers per hour, we're looking at about forty-five minutes. They may, of coarse, dissipate before reaching shore."

"And then again they may not. We've got to assume they will reach land. Would you put that region on the big screen, please?"

"Sure, here it is. We can bring up greater detail," and with that Indira put a very detailed map of the tornado-warning zone on the large viewer in the front of the control room.

Elawa interrupted with a new report. "Another aftershock in Colombia. Looks like about a five point eight so far. It's still going. The epicenter is south of the other quakes and closer to the Peruvian border. I'm sending data out to them right now."

"Any reports coming in yet?" Kristen asked Elawa.

"Nothing yet, but we do have an open link into the region. Something is starting now. It's from our planes in the area. The rain is still coming down hard, and they had to fly south to refuel, so they were on the ground when the quake hit. The airport is all right, and so is everything else in their immediate field of vision. They are going airborne in just a couple of minutes, although they don't expect to be able to see much due to the rain. They want to fly over the dams and bridges in the new quake zone to check their structural integrity and assess the need for immediate evacuation or not. It wasn't a real strong quake, but some of those dams are a hundred years old or more, and it wouldn't take much to bring them down. They'll be in touch with us as soon as they know more. Here's Paulo on the screen now."

"Headquarters, this is Paulo, do you copy? Over."

"Yes, Paulo, this is Director Verba, and I hear you loud and clear. What's happening?"

"We're still feeling the earth move here, but it seems to be subsiding. How strong was it?"

"It looks like five point eight," Kristen responded. "How are the buildings around you holding up?"

"So far, nothing else has collapsed in my view, but I think that is mostly because everything that was already weakened has already collapsed. That last aftershock before this one really took its toll on anything that was at all in a weakened state."

Paulo paused, then went on. "I'm getting reports from the ground near the epicenter, and it doesn't sound good. More buildings down, people trapped inside—most people were inside because of the rain, even though it is daytime here. Otherwise the majority of people would have most likely been outside. The rain is itself becoming a significant concern. We're getting some small stream and river flooding, and if it keeps up, we'll have flooding of the primary rivers and lakes too."

"We are working on getting more planes, boats, and other equipment to the area, and the first ones will be arriving soon,"

Kristen said. "We can start moving people out of the flood areas, and by the sounds of your reports and the computer projections for more rain in the area, we should start putting out flood warnings for all river, lake, and stream areas right away."

"We've already done it, about an hour ago, when we saw the latest rainfall estimates coming in from the computers," Paulo answered.

"Good, keep me informed. Out for now."

"Copy, I'll be speeding evacuations once those aircraft are here. Out for now."

"What are the waterspouts doing?" Kristen asked Indira as she turned away from the big screen.

"They're still on the same course and same speed. Most of where they are aiming should be evacuated by now anyway as it is in the low delta regions. Hopefully they won't injure anyone although property damage could be substantial because of how many structures there are in that part of India. We're showing at least twenty-eight waterspouts, and these blurs here in the radar could be additional ones. It is sometimes hard to tell when they get so close together."

"What is the time frame now for landfall?"

"About twelve minutes," Indira said. "I'm getting word that the landfall area has been evacuated, and they can see them from the air now. The planes are from the Indian scientific organization. They're one-person jets for weather observation, which is why they're not helping with the evacuation.

"One of the pilots is going to attempt a fly-by of the tornadoes. I'll put it on visual. What we're going to see is the live pictures from the plane's onboard cameras. It's dark, but I think with a little computer enhancement and night vision we can see them. Look, there they are." They all watched the eerie pictures coming in the storm front.

The band of waterspouts was just moving inland as the cameras caught their motion. It was strange observing a scene like this halfway around the world from where it was taking place. The waterspouts turned to tornadoes as they moved over the land. The first buildings in their path were quickly destroyed, exploding from a tornado's low pressure and the pummeling winds.

The plane banked sharply and came back for another pass. With the help of the computer enhancement on the infrared cameras, they observed the passage of the tornadoes through small coastal villages where the destruction was widespread. Fortunately the people had been gone for some time now, and no one was injured from the twisters. After a while the storms began to dissipate, and in another fifteen minutes they were all but gone.

"That was impressive. I've seen tornadoes before, but never that many at one time. They sure did a number on those buildings. It was a good thing the area was evacuated; I don't want to think how many might have been killed otherwise."

"Here's Helene again, Kristen," Indira said as Helene's picture came on the large screen on the wall. She had hooked up the visuals to watch the tornadoes herself when she heard what was going on with the weather plane.

"You all saw that?" she asked.

"Yes we did. Some line of storms, aye? Everyone all right?"

"As near as we know, there was no one in the region, save for the pilot in the observation plane. That was some fancy flying in there. I wouldn't mind having that pilot fly for the Squad. Pretty gutsy too. Anything else new that we need to know?"

"No, we've been keeping your staff supplied with up-to-the-minute information, and nothing's really changed in the direction or speed of the storm since we last talked. There was another aftershock in the Colombian earthquake, and there are reports of more damage. Other than that, you're the story right now. How's the evacuation going?"

"It's still on track, although it is getting a little harder as the rain and wind increase. We're showing gusts now to fifty kilometers per hour along the coast already, and the rain comes in waves. We're in a lull right now, but you can still see that the rain is steady. We have radar tracking of another large wave of rain that will be here in about ten minutes. What's the highest wind speed in the eye wall that you're showing right now?"

"You've had some gusts in excess of 330 kilometers per hour in the last five hours with sustained winds near 295 kilometers per hour. Those winds have continued at near that velocity for most of the last twenty-four hours. That makes this a very dangerous storm and the strongest one ever recorded. Barometric pressures are also in record territory."

"Well we'll be as ready as we can for it when it hits," Helene stated as a matter of fact.

"When are you leaving the region?"

"We're not leaving. We have some caves that we're going to move into when the full fury of this thing hits, and we're not in any flood danger as we're up on the side of this hill overlooking the river below. I'm sure we'll be all right, and besides, then we're here when it's over, and we get right back to work. Also we have a lot of scientific instruments that we want to monitor while the storm is in process. We're going to hook up some cameras too, so that you will be able to watch live as the storm hits. Anything else right now? I'm needed inside for a minute."

"No, not right now. Keep in touch though, say, in one hour again?"

"Okay. Talk to you in an hour, and don't worry, Kristen. We're in a safe place even for four-hundred-kilometer winds. Bye," she said, replying to Kristen's unspoken concern for the Squad members in the storm's path. Kristen knew that Helene was a competent leader and would never put her charges in any needless danger, but still she was concerned. She had the responsibility after all for everyone in the Squad, and with that came a lot of worrying over everyone and everything. But if Helene thought they would be safe there, then she would say no more.

"What's the latest on Colin?" Kristen asked, getting back to the situations at hand.

"Only a slight change," Markus responded. "It's moving on a slightly more northerly track with about the same wind velocities as the last twenty-four hours. Ocean surges are reaching most of the islands in the Caribbean and throughout the Bahamas. Some rough seas are starting to show up even on the coast of Florida. We have full alerts out for all the regions, and evacuations are well under way everywhere in the sector. Even if the storm picks up forward velocity, we would be ready.

"Landfall on the first islands in Colin's path is expected within eighteen hours. It is on a route that will take it just north of the most eastern Leeward Islands. Then it will likely move toward the Turks and Caicos Islands and into the southern Bahamas. From there it is projected to cross southern Florida and into the Gulf of Mexico."

"Sounds like we have done everything we can to be ready for it," Kristen replied. "Now we just have to wait. Keep me informed if anything changes—and regular updates on the hour from now on, please.

"You've got it," he replied.

"Thanks, I'll be back in an hour, but first I want to check on some business in my office. I need to keep up on my reports to our sponsors."

She went back to her office and found Silverton asleep on her desk right on the papers she needed to work on. Gently she picked up the sleeping calico and put her on her lap as she started to work. She thought that she had moved Silverton without waking her up, and although she never opened her eyes, she did, however, start to purr.

"Markus, let's run another set of computer projections on Colin now that it has turned slightly more to the north," Natalia said after the director had left the room. They went to work on the new predictions. Meanwhile Elawa was monitoring the earthquake in Colombia and the forest fires still burning in scattered locations throughout the world. Soon it would be time for Markus, Natalia, and Elawa to take their four-hour break, and all were looking forward to a couple of hours of sleep. Their relief came, and after updating them on what was happening, they went to get some well-deserved sleep. Since they were so tired, they were asleep in minutes.

After what seemed like just moments, the wake-up call came and they were on the way back to control headquarters. Kristen was standing by the Asian storm-tracking screen and was talking to Helene as they entered.

"Helene, we keep getting a broken signal. Can you switch to another frequency?"

"Is this any better?"

"Much. You're coming in fine now. We also have your video on screen too. Looks like the winds are still not too bad. How high are they now?"

"The last major gust hit ninety five kilometers per hour, and sustained winds are at about seventy. Closer to the coast we've had reports of wind gusts near 130, so down there the winds are officially hurricane strength. The rain near the coast is really coming down hard at almost three centimeters per hour. That's a lot of water if that keeps up for any length of time."

"Our satellite picture shows the storm coming onshore nearly due south of Calcutta. That puts your headquarters right in the path of eye of the storm. Are you sure you're all safe?" Kristen asked

"For sure. We have five caves into the side of a small hill well above any floodplain. They were equipped as a place for royalty to hide during war, even before the British came here hundreds of years ago, so it's actually pretty plush. True, it was looted over the centuries, but a restoration project recently has brought it back to nearly its original splendor. We were fortunate that the local authorities offered this location to set up our emergency control center. It is ideal for its central location to the soon to be affected areas. And now it seems like we're really going to be in the center of things."

"It sure does. Well, stay protected and contact us again in an hour, if you would, please."

"Will do; talk soon."

"Hi, Natalia, Markus. I trust you two had a good nap," she said after signing off from Helene.

"We did, but it was so short. Anyway, what's happening?"

"I want you two to run some projections on flooding from different rates of rainfall over the region. Project for each river drainage basin, and compile for the overall picture in the main rivers of each drainage zone. When you're done, let me know right away, please."

"We're on it," they said in unison, which brought a smile from the director.

Kristen left them on their task and went to talk to Elawa, who was also back from her rest period and was again monitoring the other events being overshadowed by the ensuing landfall of the Nisha.

"Get some good rest, I hope?" Kristen asked her.

"Indeed. Here's the rest of the world report. Those fires are still burning out of control here and here. The earthquake in Colombia has had another three minor aftershocks, but they didn't seem to cause any more damage. There has been one ship taking on water near Iceland after it struck an unidentified object just off the coast. We're working on that now and will have all the crew off in just minutes and start a salvage effort to stop the ship from taking on any more water.

"Other than those events, the world is reasonably quiet save for the two hurricanes. It almost seems like those two storms are soaking up spare energy around the world. It is mighty quiet for this time of year in areas where we usually have spring storms that traditionally trigger flooding in the Southern Hemisphere. Even the number of fires we're having right now is below the average for the last twenty-five years. I checked it out earlier just to get an idea if the number we have under way now is normal or high for this time of year. And it turns out to be low. It is fairly common after summer in the Northern Hemisphere for the timberlands to be dry and susceptible to fires from late-season dry thunderstorms, which strike quite often in North America, Europe, and Asia. We typically have twice this number or more under way worldwide this time of year."

"I guess you're right, thinking back a few years. Maybe those storms are sapping more energy than we even thought from the atmosphere. That would make an interesting study. Since you brought it up, maybe you could come up with some experiments to prove your theory one way or the other?"

"That would be great. Thanks for the opportunity. I love this kind of challenge"

"Great, make it so." Kristen left her and proceeded back to the Asian storm screen. It was showing the storm bearing down on the Indian coast with intense fury. Helene's live video of the storm in progress was feeding to the large wall viewer and was increasingly showing rain and heavy winds spreading onshore.

She stood and watched for a moment, and the view reminded her of a time in her home nation of Australia. Her home on the northeastern coast of Queensland also had storms such as this one. Although none were ever as strong as Colin or Nisha, nonetheless they were intense and did damage to the Great Barrier Reef. She had lived through many of these storms and knew what the people in Asia were feeling even as she stood there safe half a world away. Her heart went out to the children who were likely frightened and their parents who were trying to protect them from the storm.

It was an awesome sight on the viewer. Trees were beginning to bend over to the ground and the rain was now horizontal. Even though the camera was tightly secured, it too was shaking in the gusts of wind. The data from the weather station that Helene had installed on top of her hideaway was showing wind gusts to 160 kilometers per hour and sustained winds of 123. Rainfall amounts were at one and a half centimeters per hour and increasing.

Helene came back on the screen, as Kristen was just about to call her. "Helene, we're watching here, and it is really coming down hard. Are you all right there?"

"Yes, we're fine in here. Take a look around, and you'll see we're quite safe in these caves." With that she panned the camera around the space she was standing in, and it showed a well lighted room with desks full of computers and radar screens and all the other equipment needed for this kind of major operation. I have some local reports to pass on.

"The eye has passed onshore just moments ago, as I'm sure you're aware. Wind velocities are hard to determine right now as the equipment we had set in place near the coast has been destroyed, but before it went down it was showing gusts to 325 kilometers per hour and sustained winds just shy of three hundred. Rain is falling at nearly five centimeters per hour near the coast, but less the further inland one goes—for now anyway.

"We don't have any real reports of damage, but then we don't really have any reports at all yet. Most people are either out of the areas being hit hardest, or they're in hiding so they don't know what is happening outside. It'll be hours before we have any on-site reports. One thing we do know now is that a major bridge near the coast has collapsed under the wind. Other than that, we don't know what is going on out there. Do you have any more information that we don't have?" Helene asked.

"Not at the moment. Satellite pictures show the storm just moving on the coast and cloud cover extending 350 kilometers in every direction. We have some reports of damage further south down the Indian coast where the storm just brushed on its way north. Most of that destruction was caused by storm surges being pushed into the coast. Many small fishing boats were beached, and one major ocean liner, which was in dock for some repairs, has been slightly damaged and is taking on some water at the moment. Quite a few tornadoes have been spawned along the fringes of the storm, but we don't have any reports on them at this time. We'll check back with you in about an hour unless something comes up first that we need to talk to you about, okay?"

"Sounds good," Helene responded; "out for now."

Kristen went around her control room getting an update from each of the Squad members on current conditions. The reports were not good; wind was increasing; rainfall was increasing, raising the probability of major flooding; and now Nisha was starting to slow ever so little. That would only increase the rainfall. After the director gathered the current data, she realized she had not eaten all day. Before she could head out for a quick bite, she was interrupted.

"Director, come over here please," Natalia said.

"What's up?"

"It's the Atlantic storm. It appears to be picking up forward velocity. Also the barometric pressure near the center of the storm is falling again. The storm is intensifying further."

"Put out immediate advisories to all possible affected regions."

"All right, it's done," Markus interjected.

"Good. Anything else changing?"

"Not at the moment, although it seems it might be changing course just slightly too, but it's too early to tell at the moment. Wait, something is coming up on the computer now. We are showing waterspouts! A line of twisters is developing just north of the Dominican Republic moving southwest at thirty kilometers per hour. I'm sending out a warning right now, and if the waterspouts continue on their present speed and course they'll come onshore in less than half an hour."

"That doesn't give the local population much time to take shelter. Are we getting any response yet from the island?"

"Yes, they've acknowledged our alert and are taking immediate action. Fortunately that area of the island is not heavily populated and is mostly open pastureland. It will be fairly easy to get everyone to a protected area right away."

"Great. Let's hope we're this lucky all the time. Where are the waterspouts heading now?"

"They are moving in the same direction, but going a little faster. Also it appears they're pushing up a small tidal surge in front of them now. It seems to be about two meters high from radar and satellite information coming in on the computer. The band of twisters is not very wide—about eight kilometers—and the surge is directly parallel to the line of storms going almost exactly the same direction," she said.

"Keep us posted," the director said to Natalia, "and call me over here when they're a couple of minutes from landfall, okay?"

"Consider it done," she responded as Kristen went back to the Asian storm screens.

Now Helene was calling headquarters. "Kristen is that you?"

"Yes it is, go ahead."

"The winds have really kicked up here where we are. You can see our wind gauge is reaching the 270 kilometer range, and near the eye we are getting reports of much higher gusts. Some of the gauges we set up prior to the storm have been damaged and are no longer sending information, but before they broke they were showing gusts to 379 kilometers per hour. I suspect that in some areas they are reaching nearly four hundred kilometers an hour.

"And the rain is nearing five centimeters per hour even here. Several of the smaller streams and rivers are already approaching flood levels, and we have reports of three bridges and two dams already broken. Also, remember we're not getting very many reports yet. Most places are still under the full fury of the typhoon and are in shelters.

"I'm sending a picture from our storm cam, and you can see the horizontal rain and how badly the camera is bouncing in the wind. Believe me, we did our best to secure it, and it's still moving that much. Gives you an idea of what it's like here right now. I'm shifting sound to outside so you can hear it too."

There was a huge roar, and Markus quickly lowered the volume on the control room speakers. All the personnel in the room flinched at the sound of the wind and the rain pounding halfway around the world. It was an awesome sound and quite impressive in its decibels. The director signaled to lower the sound level even more, and it became much quieter in the room. It was an eerie quiet after the full thundering noise of typhoon Nisha.

Helene came back online in another moment. "Well, what did you think?"

"That was some sound. How are the coastal areas doing with the tidal surges coming onshore? Do you have any reports yet from those areas?"

"Not yet, there aren't very many people in those regions. In fact, in some of the coastal cities there is no one at all. Others have only a few people at most. We were pretty thorough in our evacuation effort," Helene said. Meanwhile, Natalia was waving her arms, signaling to the director that the waterspouts in the Dominican Republic were coming onshore.

"Helene, I've got to go. We have some twisters hitting land in the Atlantic storm right now. I'll call you back in a bit after these storms are played out. Talk to you soon." Kristen moved back to the screen where Natalia was monitoring the line of tornadoes just coming onshore.

"Where are they now? Are they posing any threat to populated areas?"

"They're coming on shore right here, and as you can see there is one small town to the east, about two kilometers from landfall and another one to the southwest—here. If the storms stay on their present course, they'll miss both villages. But there isn't much room for error. If they move at all either of the towns could be in their path. We've sent word, of course, to those two villages and they went to whatever shelter was available on just a moment's notice. Hopefully they'll be all right, and the twisters will stay on the open pasture line in between the two towns."

"Let's keep our fingers crossed," Kristen said, as they all stared intently at the monitor. A few of the tornadoes began to dissipate as they moved further inland.

"Look!" Natalia said. "These three have shifted right toward the town to the east. Impact in five minutes. I'm sending the information to them … now."

"Any response?"

"Not yet, but it was received. They probably don't have time to respond just now."

"I bet you're right. With the storms so close, they'd better be taking—"

"Wow," Natália broke in, "we have been able to link into their town cam, and you can see the tornadoes heading right at them. This is some sight." She cast a sheepish look at Kristen. "Didn't mean to interrupt you, but I guess I was a little overeager for everyone to see these pictures."

"That's okay," Kristen said. "I would have done the same, I'm sure. These are truly macabre scenes to be watching in real time and not know the outcome." The room became quiet with only the roar of the twisters growing louder.

"Impact in three minutes … two minutes fifty seconds … two minutes forty … two minutes thirty." Natalia stopped the verbal countdown as the pictures coming through told the rest of the story. The next two minutes had every Global Disaster Squad member in the control room staring at the big screen where the pictures from the Caribbean had been transferred. It became apparent that one of the tornadoes was going to squarely hit the camera, which was located on the edge of the small town.

"How close are our people to this town?" Kristen asked Natalia, knowing that when the twisters were sighted, Squad members had begun heading toward the possible disaster area.

"They're just minutes behind the line. They're following the twisters so they can start rescue procedures immediately if needed. It looks like they might be needed as the town is right behind these cameras." As she finished speaking the tornado was only thirty meters away. They all watched as the storm moved on a straight path at the now violently shaking camera. The screen became filled with the twister, which was not all that large yet still packed a powerful punch as it slammed into the mounted video camera. The screen went blank. The sound ceased. The room was completely silent.

"Are you still in contact with the island?" Kristen asked Natalia, who was busily trying to raise the island.

"I've lost them. Switching to Squad members. Control to Global Disaster Squad unit in Dominican Republic. Do you read?"

"We copy. We're following the storms into the town. Can't see much yet, the rain is very heavy, and visibility is very short. We are beginning to spot some damage. Buildings strewn about, shattered into millions of pieces. Trees uprooted, automobiles overturned. Stand by and I'll connect our monitor to you."

In a few moments the screen again filled with the images coming from the Caribbean. Again it was the picture of horizontal rain, dark swirling clouds, and glimpses of the distant line of tornadoes the Squad was pursuing.

"We copy the video. Where are the twisters?" the director asked the pilot.

"The town is relatively small, and they are moving toward the far side already. It looks like two or maybe three actually went through the town. The others have moved to either side and did very little structural damage, although some trees and livestock may have been destroyed along with planted crops and orchards. We're right behind the ones leaving town, and some them are beginning to break up. There's only about half as many as there were when they came onshore.

"They're moving into open land for now, and we're circling back to look for anyone needing medical attention. As you can see, now as the rain lets up, there are three definite lines of destruction through the town. The twisters are not very wide, which was fortunate, but where they did go, there's not much left. I see some people who may need help, and we're setting down now. We'll leave our video feed on, but we need to sign off for a moment and check out the rescue efforts. Over for now, we'll call back with a report as soon as we have it."

"Thanks, talk to you soon," Kristen responded, still watching the screen as the chopper sat down amid the devastation. They watched a moment longer, and then Kristen was called over to the Asian storm monitor.

"Director, it's Helene. She's back on the line."

"Kristen, we've just learned of a major dam that is giving way on one of the tributaries of the Ganges. Wait, I'm getting a message now … It has broken, and a wall of water is heading downstream. We're sending an emergency alert now. This could be real bad, as some of the villages along the path didn't completely evacuate, since they're high enough to avoid coastal flooding. We are getting more … The first village downstream was just completely washed away."

"Do you have the number of villages in the flood's path to the Ganges?" Kristen asked.

"Not for sure, but at least eight that definitely are in the path. We have our people on the way, but it has to be by land as it's too dangerous to fly right now, and that will take a while. The roads are washed out in many places, and most are covered with at least some water. We're using the all-weather vehicles, but there are stalled vehicles and fallen trees everywhere, so it will be slow going. It will take at least a few hours. We'll keep you informed.

"Here's the rundown on the rest of what we're hearing," Helene continued, giving the details of the situation in India to the control room. It was mostly the same across the region. Heavy rains, strong winds, flooding in many areas, communication out in nearly half the area, but fortunately not many injuries or casualties were being reported. Kristen thought that those numbers might be changing when the damage from the dam breach was reported. The screen in the control room was switched to the India feed, and a picture of Helene in the emergency center came on.

"Can you see us?" she asked Kristen.

"Yes, we have you."

"Good. We're going to show you some of pictures we're getting. The first one is from the coastal area, and you can see the rain is still coming down at a steady rate. The winds have subsided only a little. They're still at nearly 250 kilometers an hour. Now we're switching to downtown Calcutta. Most buildings were boarded up, but many are still showing damage. The rain is coming down in sheets and moving nearly horizontal. The winds here are at about 210 kilometers, which is plenty to cause a lot of damage."

"Helene, the picture from Calcutta is breaking up, can you switch to another Calcutta camera?"

"Sure, we've got several to choose from in the city. Here's another one. Is this coming in okay?"

"Yes, thanks. What area are we viewing?"

"Near the train station. A lot of damage can be seen between the rain blackouts. Most of it is minor, although I can see some buildings with structural damage. This may change if these winds keep pounding them for much longer. Now we're switching to another view from near the airport. Pretty much the same picture here.

"Let's move further west to the city of Kharagpur. More structural damage to the buildings. Many trees uprooted, vehicles tossed around like toys. Now back east to the capital of Bangladesh, Dhaka. Here you can see that the rain is not as heavy and the winds are less than two hundred kilometers per hour. It looks like the strongest part of the storm is from the border of India and Bangladesh westward to about the Garhjat Hills in Orissa state. This region is getting the worst of the storm. We do have a feed from Orissa near the coast from the city of Baleshwar.

"Wow! It's almost completely underwater. You can just see the tops of some trees and a few buildings. This was one of the places where we had complete evacuation because it is so close to the water," Helene said, obviously overcome by the scene in the coastal city.

The camera panned the scene, making a complete 360-degree turn. It was the same in every direction, a few trees and a couple of buildings peeking through the water. The rain was abating somewhat, and the winds were beginning to subside, but the damage was already done. The entire city was underwater.

The scene shifted to another small village farther inland where the rain was coming down in what appeared to be buckets. This village was in the hills, and the higher elevation was wringing every possible drop of water out of the storm. It appeared that the village was not flooding that badly, as it was on the side of the hill, but the water running down the mountain was becoming a small river, now cutting a gully toward the lowlands. The flooding downstream was going to be intense.

"Helene, do you hear us?"

"Yes, go ahead."

"That looks really bad in the hills. Which river will that be flowing into, and do they know what's coming?"

"Yes, we've sent word out already to all downstream areas to seek higher ground immediately. There are many little streams coming out of the hills, but they all reach the Sone River or the Budhabalanga River. The Sone meets the Budhabalanga near the city of Baleshwar just northwest of the city. That was the city we saw underwater, so it can't really do much more damage than has already been done. It is the little villages on the way downstream that I'm concerned about now.

"And it appears that communication is cut off everywhere in the area. We can't get any planes or helicopters off the ground right now, as visibility is nearly zero in all this wind and rain. I just hope they got our warnings before communication ceased. Of course even if they did get our message, it's going to be hard for them to get anywhere in a hurry, because the roads are most likely flooded, and there are bridges out all over the place. Any suggestions?"

"How much time do you figure before the main water gets to the villages?"

"Maybe three to four hours."

"What if they tried to go by river now before the water gets any higher? If they could reach some of the small hills that are downstream and climb to safety, there's a chance that many could be saved."

"It might work, but how do we get the plan through?"

"Keep trying regular lines of communication. Maybe some radios are working, and I think we've got to send some sort of aircraft into the region. How many seaplanes are in the area?"

"Maybe seven or eight that I know of; they're not real common in this part of the world. Are you thinking of landing on the river?"

"You got it. What do you think?"

"It just might work. Plus we could get some of the people out on the planes. The problem will be to try and navigate in these conditions."

"Are they equipped with satellite sensors?" She checked her information on the planes and found a few that were equipped to fly by GPS satellite guidance.

"Yes! That should work."

"Here's the frequency that we'll be broadcasting on. Get this to the planes and we'll start the automatic tracking of all of them as soon as they're airborne."

"We're on it now. Let's hope this works. There's not much else that we can do as long as the weather is this intense."

"Well then, it's showtime!" Kristen felt the adrenaline starting to pump in her body. It was always hard to ask others to put their lives on the line, but it went with the territory and she did her job well. After all, it was their job to save as many lives as possible, and this was just routine Squad work. The planes were in the air in less than five minutes from when the order went out. They were just out of the main brunt of the storm, so they were only anchored slightly, which made short work of getting off the ground. The navigators tuned in to the information coming from above, which was the result of sophisticated radar detection equipment left over from the days of military spying by nation states. It had been modified for more peaceful purposes now and worked primarily as a navigation beacon to planes and ships around the world. It could, when required, be used to track the positions of individual craft. Again this was used for the most part when a vessel had lost its own inboard radar equipment or was in an intense weather situation, as was the case now. The planes were picked up by the satellite as soon as they were off the ground, and location data began to flow directly into their navigational computers, which had their course already laid in to the targeted area.

"The planes are airborne and on their way," Helene stated over the video hookup.

"We've got them on the satellites already. We think they'll be safest flying south first and then coming back in from the south, as the storm is moving north. Winds are diminishing over the Bay of Bengal where we are routing them. It will add about half an hour to the trip, but it will give them a much better chance of a successful mission."

"Sounds good to us here. We instructed the pilots to try and land closer to the banks of the river, as the water turbulence will be less than the middle zone. Plus that puts them closer to the people they're trying to warn about the wall of water coming down from the hills. Most of the people living along the river have boats, so the majority of them can hopefully get to higher ground before the water really rises. We've located some hills along the river where we've projected that, if things stay at the present conditions, they'll be out of danger. As many of the older people and others who will have a hard time getting to the higher ground will be airlifted out of the area. However, we won't have room for too many and we'll only get one chance at getting anyone out, as the water will be there before we can make a round trip."

"Understood. How do you propose to get the message out along the riverbanks?"

"The planes were loaded with loudspeakers, and we have a taped message in the local dialect, which we'll play continuously as soon as they're in range of the villages. Then we're counting on the locals to pass the word among themselves to more outlying areas, the hard of hearing, and anyone else who doesn't hear the message for themselves."

"Let's hope this works!" Kristen answered. They would have about an hour before the planes would be approaching the villages along the flood path. There wasn't much else to do now but wait for the planes to get to their destination.

Kristen told Helene to call her when the planes were about to land and went back to the screen monitoring the situation in the Dominican Republic. Natalia was still watching the feed coming in from the Caribbean island, while the others had been observing the situation in India.

"Anything new to report?" the director asked.

"The first casualties have been reported. At least seven confirmed dead, over two hundred injured, and a lot of people still unaccounted for at present. There were quite a few buildings destroyed, and many of them are just in big heaps of debris. Most of them housed people when the twisters came through. So there is a strong likelihood more victims will be found in those buildings or rather what's left of them.

"In the area right outside of town there were several farms that were completely leveled. We have some reports of dead cattle and horses in those regions also. The good news is that the twisters have dissipated and are no longer a threat to anything. After they left the villages, they didn't last long at all—maybe another eight to ten minutes at the most. And in that time they did only minor damage to some citrus groves and some banana plantations.

"The rain has let up enough to get the rescue operations under way with relative ease. But now we have reports of squalls headed toward the town that had the tornadoes. Wait—here are more reports coming through. Another nine bodies have been unburied. More injured have reported in for treatment. And here's some good news: twenty-three were found alive and uninjured beneath one of the collapsed buildings. It seems they hid inside a walk-in refrigerator unit on the ground floor of the building, which, when it went down, just piled up on the unit. They were a little cold, but otherwise all right."

"That's good. We need some more reports like that. Anything else? Are all the Squad members okay?"

"We had two Squad members slightly injured when the building they were in collapsed on them while they were rescuing a survivor," Natalia said. "They did get the person out with only a couple of extra cuts and bruises than they already had. The Squad personnel had mostly cuts and some bruises themselves, although one of them did suffer a broken wrist when it got caught between two falling beams. She has been treated and is back on duty helping other injured at the hospital, which the twisters fortunately missed. That's about all there is to report at this time, Director."

"What's the latest on the hurricane itself?"

"Speed and direction the same, although the barometric pressure is still dropping, though slowly. It is very likely that the winds will increase in velocity as the pressure creeps down more. The pressure is approaching the all-time record low. Evacuations along the Gold Coast of Florida are in full swing. Some of those people have lived through some of the most destructive storms on record for the area like Hurricane Andrew back in '92 or that other one in, was it 2008 or 2009, anyway you remember the one, I'm sure."

"I sure do," Kristen said. "It was when I was a kid and my family was visiting Disney World in Orlando. The storm went south through the Miami area, but we did get some of the effects up in Orlando, as it was quite a massive storm. What I remember, being a kid, of course, is that we lost three days of being able to go to the attractions. We had to stay inside in the hotel and watch movies on the television in our rooms. My parents extended their stay by two days though, so we made up some of the lost time. I've got a mental block on the name of that storm right now, but I'm sure I'll recall it later. Probably in the middle of the night. That's when I remember most things that I forget in the daytime. Yeah, if they lived through some of those storms I'm sure they do have full respect for Mother

Nature and are very happy to get out. When will the majority of the people be out of the area?"

"Right now the schedule is for about twelve hours ahead of the first gale-force winds. The storm is still about thirty-six to forty hours away from landfall in Florida. Gale-force winds are extending about three hundred kilometers from the eye right now, so maybe about fifteen to twenty hours before that winds will start to reach gale-force velocity at the coast—about sixteen to twenty-four hours from now."

Natalia went on, "The southern Bahamas are already picking up gale-force winds, and so are the northern coasts of all of the islands within that zone south of the storm. Not much damage to report at this time, but we expect that to change as the full fury of the hurricane comes to bear on the islands in its path. There is a high probability of severe tree damage and an almost as high chance of buildings being destroyed. Many of the buildings in the area have been built to withstand fairly high winds, but I'm afraid that the winds in this storm are likely to exceed their limitations. That is based on current conditions, which we also project to continue, if not intensify even further, if the pressure continues to drop."

"Make sure everyone in Colin's path is fully aware of the intensity of this hurricane, and it might be helpful to compare this one on the same level of ones you just mentioned. It very definitely is a category-five storm. How far north are evacuations being carried out?"

"Along the coast, all the way to Jacksonville on the Atlantic side and north to Pensacola on the Gulf side. Inland people are taking precautions as far north as Orlando. To the south the entire northern part of Cuba is evacuating, as well as most of the Bahamas. After that the entire area bordering the Gulf of Mexico is on alert. They're all waiting to see where the storm goes after it crosses Florida. The projections put it near Houston at the moment. But that can change after the hurricane hits the warmer waters of the Gulf. Storms have done everything in the past from turning around and going back into

Florida to heading into Mexico. So it's hard for now to say anything too definitive."

"Keep me posted. We all need to get some rest too before Colin hits land. Why don't you and Markus take a six-hour break and let Susan watch this station until you get back? When you two return, she and Elawa can take their breaks. But stay in the building; we may need you if things change without warning."

"We will. I really could use a break. We'll see you in six hours. Later."

"Bye," Kristen said as they left. She then instructed Susan to take over monitoring the Atlantic situation and crossed to Elawa, who was monitoring the rest of the world.

"How's everything else going?"

"Not too bad. We had a report of a high-rise fire in Tokyo. Squad members helped with the rescue on the higher floors. They used helicopters and got just about everyone out safely. There were some reports of a few injured, mostly due to smoke inhalation.

"Speaking of fires, three of the forest fires we've been fighting are just about under control if the winds stay calm. Two of the fires are under control, one in North America and the one in Southern Africa, but the other three, one in Northern Asia, the one in Europe, and the other one in North America, are more out of control than ever. Winds in those areas are whipping the flames up in the dry timber. Heavy spring rains in all three of the regions created a lot of underbrush, which is now so dry that the slightest spark sets it off immediately.

"The other problem we're watching, of course, is the Colombian earthquake. Two more earthen dams have given way, now bringing the total to somewhere around fifteen. We don't have confirming data yet on some the dams, and a few may be different reports on the same dam. Communication is out in most areas, and the rain is still falling quite heavily in much of the region. There have been hundreds of deaths confirmed and thousands injured.

"These people are going to require massive amounts of aid to help them rebuild and just to survive for the short haul. Most water supplies throughout region have been contaminated, and bottled water will have to be sent in until treatment plants can be restored to full operating capacity. We have just about everyone from the Squad in South America working on this right now, although a few have been shifted to the Caribbean to assist with evacuation efforts on the islands."

Elawa surveyed her screens. "Then there was one more report of a small plane, which went down in the Yukon, and we're assisting with search efforts in that rescue. There were five passengers and two crewmembers on board the plane. They disappeared while approaching the Alaskan border. That was about four hours ago, and the first search planes are now in the area where the plane was last known to be. Otherwise the world is fairly quiet. As if that isn't enough going on, eh?"

"You got that right. Well, it looks like we're doing all we can be doing. Can we get any help to Colombia from Central America or Western North America?"

"Central America has already sent almost every Squad member on active duty in the region already, and North America has sent some, but most of the Squad members there are either on their way to India or the Southern United States waiting for Hurricane Colin to come ashore and assisting with evacuation efforts."

"We need more help in Colombia. We need to activate a thousand reserves to assist with that effort at once. Pull half in the medical fields and the others from rescue and support teams. We can't really spare any more from North America with the hurricane approaching within thirty-some hours, and I'm afraid we're going to need every one of them when this storms comes onshore. We must have a few thousand more reserves in South America we can still call on and at least a few hundred in Mexico and Central America. Can you give me a count as soon as you have the data, please?"

"Sure, it should only take a minute. I'm accessing the main computer on personnel right now. Here's the report coming online. I'll feed it to that screen right in front of you. Here it is now." Just as Kristen sat down to view the monitor, her little friend Silverton, who was feeling ignored by all the goings-on, jumped on her lap and demanded to be petted. The director automatically started to scratch her behind her ears and stared at the screen as data began coming through.

"Thanks." Kristen was close to her guess on how many reservists were in the area, but then again it was her job to know. She reviewed the lists of available Squad members. "Let's call everyone for a limited two-week tour of duty. And let's do the same throughout the entire Western Hemisphere. I bet we're going to need them before this is all over, plus something else is bound to happen while these other situations are in full progress."

Elawa sent the alert out immediately, and in just moments the first reservists were reporting in for their active stint.

This was the first time that every Squad member in a hemisphere had been activated at the same time. If anything else occurred in Asia, Kristen reflected, they would have to call in some more help in that part of the world too. The quakes in China had subsided for now, but there was always the possibility of further aftershocks. And those pesky forest fires were still a major concern; plus there was a good probability of more fires if there was no measurable precipitation soon in those areas, and none was in the immediate forecast. Already almost every firefighting unit in the Squad worldwide was on active duty or at the very least standby status ready to go at a moment's notice.

"Reservists are checking in everywhere, Director", Elawa said after a few minutes. "They're showing up at bases throughout the regions, and we will have the first groups in route toward Colombia within the hour."

"Great. Keep me posted. I'll be back in a bit. By the way, when Natalia and Markus come back, you can take a six-hour rest break along with Susan, but like I told them, please stay in the building in case you're needed right away, okay?"

"You got it. And thanks; I'm going to need a rest by then. I'll keep you posted. By the way, how's it going in India?"

"I'm on my way to find out right now," Kristen said. "The last I knew there were dams breaking, heavy flooding, tornadoes and so forth. But the loss of life, as far as we know for now, has been kept to a minimum considering the intensity of the storm. That is due in large part to the evacuation effort prior to the storm's arrival. But it's not over yet, and we still have some of the worst to come when all the rivers reach maximum flood stages. I'll fill you in when I come back in about half an hour."

"Thanks, see you in a bit."

"Any word from Helene yet on the arrival of the planes?" Kristen asked as she approached the Indian storm monitoring station.

"Not yet, but by our calculations they should be arriving there within ten minutes."

"See if you can raise her."

"Too late, she's coming on line right now. Here she is now." And with that the screen returned to the Asian continent showing Helene standing inside the Squad's Indian headquarters.

"Kristen, is that you?"

"Yes, we're copying you loud and clear. Go ahead. Are the planes there yet?"

"They're five minutes from landing, but the waters are already higher than we were predicting, making this a little more risky than we previously guessed, so I have some serious second thoughts about all of this. They're all willing to go in, and I'm willing to support it if there is a decent chance of success, but I'm not willing to lose good Squad members when you know we're going to need them even more after this is over. Good pilots are hard to come by—"

"I read you," Kristen said interrupting Helene. "What if we send in only one or two and have them blast their recorded message and then split without even landing?"

"That might work. The problem is that the river is much more turbulent than we thought it would be by now. It looks like a lot more rain has fallen in the drainage area than we have monitored so far, and the river is about two meters higher than we had calculated. I think we should take two volunteers and have the rest turn back. If we lose these planes now, it could be very costly later on if we need to get people out of somewhere in a hurry. Do you concur?"

"Make it so."

Helene asked the crews to volunteer and then made what she hoped was the best choice: she picked two planes to continue, and the rest turned back. She picked the pilots with the most experience in heavy-weather flying and relayed them their new instructions. They acknowledged and turned on their forward videos for everyone to watch as they approached their target areas.

The planes began a gradual decent into a river basin now fully swollen with water right up to its flood level. Many inhabitants were already escaping to their boats as the water continued to rise. The planes began broadcasting, and by the waves of the people they could see through the wind-driven rain, they had every reason to believe that at least some them understood the message.

The planes made a pass down one bank, rose into the air, made a sharp turn, and came back down on the other bank, playing the recording all the time. Again they were optimistic that the broadcast was being heard and understood. They repeated the entire operation and then were instructed to move up the river to see if they could spot the wall of water approaching the villages. It didn't take them long before they spotted it; it was coming!

"We see it. Do you copy our video?"

"It's a little dark, but we do copy. How far upriver are you, and can you project how fast is it moving?"

"We're about one hundred kilometers from the village where we last broadcast the warning, and the crest of water is traveling around sixty kilometers per hour, so we've got about one and a half hours before it reaches the first of the villages. Orders?"

"Go back to the villages, and try to convey the time factor to the people there. It looked too dangerous to land, so after you have spread the word, come on back. We'll try to get some choppers into the region to help with the rescue effort after the weather subsides enough. Do you have enough fuel for a safe return?"

"We do. We could attempt a landing, but I'm not sure we could get up again safely. The waters are pretty turbulent. We'll be back in about an hour and a half. Over."

"We copy you back in one and a half hours. Good luck."

"Thanks, out."

Kristen spoke up. "Helene, we copied the first part and the last, but we lost contact when they were over the wall of water. Can you fill us in, please?" Helene filled in the details. They made further plans on the next moves to take and set a time to be in contact again unless something changed unexpectedly.

Kristen was tired. She had been at the headquarters since … well, since the party at Susan's, and except for a few minutes she had been awake this entire time. She needed to get some rest before Colin hit the mainland of North America, so she left with Indira in charge.

She knew that the next several days would be demanding, exhausting, and would take all the energy she could muster. And she would muster whatever would be needed. After all, she was the director of the Global Disaster Squad. She went to her office and on the cot in the back of her office she quickly fell asleep. Silverton crawled up on the cot with the director and soon was purring in her own dreams.

CHAPTER TEN

September 10, 2035

Kristen awoke with a start. She had been dreaming about her home in Australia, and her parents were needing her help. She thought that perhaps she should call them immediately, but she hesitated; maybe she was overreacting. She called.

"Mom, is everything all right?" she asked when her mother answered the call. Her mother started to cry. Her brother, Kristen's uncle, had just had a stroke two hours before.

"How did you know?" she asked.

"I don't know," Kristen replied, "I just knew." They talked for another half-hour until a call came from the control room. It was Indira. Colin was bearing down on the Bahamas. The storm was only a few hours out now, and already hurricane-force winds were hitting the outer islands. It had begun.

"Mom, I'll call back later. The storm in the Atlantic is moving into populated areas, and I'm needed in the control room. Give Dad my love, and I hope Uncle James gets better soon. Bye." And in less than a minute she was walking into the Squad's control room. Her little friend was right behind her and, upon entering the control room, took her usual position on top of one of the computers where it was warm, yet near the action.

Pictures from the outer islands of the Bahamas filled the wall screen and, as in the Indian storm, palm trees bent under the fury of the wind while horizontal rain dimmed visibility.

"Director, hope you had a good rest," Indira said as the director entered the room. "Here's the most current data on Colin." She continued showing Kristen the wind speeds, precipitation amounts, forward velocity of the storm, and so forth; Kristen tried to pay attention, but her mind kept wandering back to her uncle in Australia. Would he live? Would he fully recover if he did live? How would Mom take it? Mom was very close to her brother, and they had lived their lives only a few kilometers apart. She made a note to call her mother as soon as she could take a break and forced her concentration back to the issue at hand. There were lives at stake, and she needed to save as many of them as possible.

She studied the news coming from both hurricanes. "What level warning are we at in the Atlantic?"

"Highest level now for all of the Bahamas, Cuba, Florida, and the rest of the southeastern United States north to Georgia, and the entire Gulf of Mexico. The computer shows the storm passing through Florida, and if it continues on its present path, it will hit the Houston area of Texas. But most of the time storms change course once they have passed into the Gulf. It will be a few days before we can predict anything; till then it's contingency planning and guesswork."

"Is there a path taken more often than most?"

"Probabilities would put the storm turning more northerly once it enters the Gulf, but just barely. However, it could completely reverse itself and go back into Florida from the Gulf side—or it could turn further south into Mexico and then come back out over the water and head into the southern United States. Here are the paths of the last hundred years of hurricanes in this region. You can see as I overlay each new storm track that rarely do two storms approximate the same course, and in fact, we have no record that any two have ever duplicated tracks. Sure, some have come close here and there, but never exactly the same."

"I see. Well, then it is completely prudent to put the entire area on highest-level alert. Now what about the other crises going on in the world?" she said and walked over to Elawa who was still monitoring the rest of the globe.

"Hi, Director. You look worried; is everything alright?"

"Not really; my uncle had a stroke, but he seems fine for the moment. That's about all I know for now." Elawa was very empathic, and it was not uncommon for her to feel the troubles, the joys, and the basic emotions of others. People sensed her special gift and were always asking for advice. With people she knew well, like the director, she would sometimes make a comment, especially when she knew Kristen would try to be stoic and not reveal her sadness or concern.

"I'm sorry to hear that. Let's hope that he recovers fully."

"Thanks, I appreciate it. Anyway, how's the Colombian situation?"

"Here's the latest from the quake area: casualties are running ever higher. We now have reports of at least two thousand dead, forty-five hundred injured, and as many as ten thousand missing. Eighteen dams are destroyed or damaged to the point that water is not being stored, and they are leaking badly. Rescue efforts are being severely hampered by the storms across the region. And the bad news on that is the storm off the coast of South America keeps sending wave after wave of squalls into the jungle-covered mountains. The first reservists have arrived in the area and are already giving assistance to the Squad members previously on duty."

"Other than that, not much new to report. No new dams have broken that we are aware of, and the rain is steady and not increasing for right now, although it is still quite heavy. Visibility is down to several hundred meters at best in some regions to virtually zero in the zones around the mountain peaks. The rains are forecast to continue for at least the next couple of days, even if there could be periods in between the squall lines that are almost clear. We can monitor those areas from space so rescue efforts can be timed to go full force in those interludes from the rain. It's amazing that this area has suffered from a drought for the last, what, two to three years, and now when we don't want rain, it's pouring."

"It is staggering the amount of rain that has fallen since the drought broke just two days ago," Kristen said. "Anyway, your idea sounds like a good plan. How fast are the squalls moving at the moment?"

"Twenty-five to thirty kilometers per hour. They're clipping along at a pretty good pace, and they're lasting a few hours with a couple hour break in between."

"Then we should be able to get quite a bit done in the few hours that we'll have. Will you relay the information to the on-site control headquarters and get them an automatic feed from the weather satellites on their region? Then they can monitor the squalls with the least delay for action."

"It's on its way, Director. Anything else changes, I'll let you know immediately."

"Thanks, and thanks again for your concern for my uncle." With an understanding glance from Elawa, the director left the control room. Silverton followed her out of the room and down the hall. She seemed to know that Kristen was worried and was sticking close to her today. The situation was under control, and she wanted to contact her mother before the main part of Colin came onshore in North America.

She went to her office and made the call. There was no real change in her uncle's condition; he was resting at present and had been for most of the day. She sensed her mother's deep concern and wished she could be there to help comfort her.

Then she cleaned up a little and went back to the control room. Natalia and Markus would be back soon, if they weren't already there, and she wanted to update them on the current situation. They were already at their stations when she entered the control room.

"You two get some rest?" Markus and Natalia were huddled over a computer screen watching the newest data coming in from southern Florida and the Bahamas, where the storm was beginning to do considerable damage.

"Yes we fell right to sleep and slept until just minutes before we came back here. We're getting updated on what happened the last six hours", Natalia said and continued, "looks like things are really beginning to heat up. This is the most current data we've got right here on the Atlantic storm." Winds were well beyond hurricane strength in the Bahamas by now, and the tides were ripping the beaches apart. Rainfall was very intense, falling at a rate of over five centimeters per hour. It was all moving toward the mainland at fourteen kilometers per hour. The storm had slowed its forward momentum, which was allowing it to strengthen even more. "So you activated the entire Western Hemisphere's reserves, I see."

"Yes, with the hurricane, the earthquake and flooding in Colombia, fires in the western United States and Canada, and who knows what might happen next, it seemed like we're going to need everyone we can possibly get."

"I see what you mean. Any project you want us to work on specifically?" Natalia asked.

"First I want to update you on both storms, and then, yes, I would like you to work up the most current computer projections on the Atlantic's storm and flood-level probabilities in Asia. Do the Asian one first, so when I talk with Helene next, I can give her that information. We are scheduled to hook up with her again in about thirty-five minutes, so you've got until then."

In a few moments Kristen had them updated on the three major natural disasters going on in the world. Then, leaving them to their assigned tasks, she went back to her office to try to catch up on other work or at least to put it in some order of priority to accomplish when these storms had played out. Her feline friend was still right with her and slept right on her desk as she worked, purring as usual.

The thirty minutes went fast, and soon she was back in the control room waiting for the linkup with Helene in India. There was a slight delay in establishing communications with India, and Kristen's mind wandered back to her mother, how she and her uncle were doing, and what she herself might be doing to help if she was there with them. But she didn't have very much time for an answer as Helene's picture appeared on the screen.

"How's it going?" she asked. "Are you all still all right in your control room?"

"Yes, we're all fine here, but we've lost contact with a few of our people. One of the planes warning the river inhabitants failed to come back. The other plane reported losing radio contact with them and that they disappeared from the radar. We don't have any confirmation of what has happened, but it looks like their plane went down. We can't get any rescue planes in the air right now as the storm has intensified in the region where they were last spotted, but we will as soon as we possibly can. Meanwhile we just have to hope that if they crash-landed, they were able to survive.

"That's about all I can tell you at the moment on that situation. The flooding is increasing throughout the area. The rainfall has exceeded all projections, and given a few dams breaking here and there, we've got ourselves a real mess."

"I copy," Kristen said; "can we assist in any way?"

"Not until the worst of the storm has passed. We just can't risk any more lives at this point until the chance of success and survival increase significantly. Do have any projections on when this baby might be ending?"

"Maybe eight to twelve more hours or so, by the looks of things right now. From here it looks like you are out of the worst of it. Then, if we're lucky, the coastal areas will be safe to fly in again. Do you copy me?"

"Ten-four. I think our best move is to lie low until it's safe to venture out again."

"I agree. Get some rest now for a few hours. We're going to need you refreshed when this all breaks and the real serious flooding and the rescue effort begin. We'll call you when we see the main part of the storm pass the coastal areas. The problem is that the typhoon has slowed down its forward velocity and just isn't moving the way it was even thirty minutes ago. So if anything changes, we'll call when you need to get ready or if the situation transforms in some major way. Until then, pleasant dreams. For now, out."

"Thanks, talk to you later."

Turning to the computer that was monitoring the Asian storm and addressing the Squad member watching the typhoon, Kristen asked for updates every fifteen minutes until it was again safe to fly into the area where the plane had gone down. Before leaving she instructed the maximum number of people to get some sleep too. After all in just a few hours the… well the shit was going to hit the fan, so to speak and she wanted the Squad to be ready for maximum efficiency when it was needed.

Then she left the control room, followed by her friend Silverton, and went to get a few moments rest herself. She knew the worst was still to come, and this might be the last chance she had for a long while to get some well-deserved sleep.

Markus and Natalia went to a resting room and crawled on the same small cot at the very end of the room, which was mostly in the dark and with no one near them for at least ten cots over. It was a medium sized room holding only about thirty small cots, all on one side of the room. They settled into an intense kiss and fell asleep still holding each other.

Elawa also went to the resting room and, thinking about the spirit of the Mother of the Earth, asked that all might be spared any more attacks on lives of the planet. She received no immediate reply but sensed that the Mother was disappointed in the way humans were handling the planet. She was in general correct about her feelings. There was something going on, but she couldn't quite put her finger on what it was she was feeling. Exhausted, she also fell asleep and dreamed of Africa and her family.

Susan went into the darkened resting room thinking about the time that she was home in southern Florida and going through the same situation that was just about to happen there again. She was lonely too and fell asleep wondering when and where her true love might be found. The others also rested, every one switching off so that the entire staff got some shut-eye.

CHAPTER ELEVEN

Four hours passed without any change in the situation. The group slept for the better part of that time. Markus had gone in once to the control room to see if they were needed and, after assessing the situation, went back to sleep when it was apparent there wasn't anything he could do to help the situation. Natalia slept through the entire four hours without so much as rolling over. She had been truly exhausted. Susan on the other hand slept very lightly, waking several times and tossing and turning a great deal. Elawa slept peacefully.

Then in a flash the rest period was over, and moments later they were all in the control room.

"It looks like the storm has subsided enough to launch a search and rescue on the missing plane," the director was saying as they entered the room. "It's still pretty choppy, but if we don't try now, we may not get another chance due to severe flooding in the region where they were last reported. As you can see from these infrared satellite pictures, the water is rising rapidly in the entire coastal region and along the river where they were last known to be. We have two planes in the air already and expect to have three more within ten minutes. Markus, you can monitor the planes and feed any satellite information to them as required." She went on to give everyone a task to assist the rescue attempt.

Departing for a few minutes, she went back to her office and tackled a little of the paperwork piling up on her desk. It seemed like weeks that they been monitoring these storms. The Indian Ocean one was starting to subside, but the aftermath would likely be more devastating than the storm itself as all the rain that had fallen in the mountains flowed toward the coastal cities, and the Atlantic storm was still to come. It would indeed be a long week.

She then busied her mind with financial reports for the Squad and funding projections needed to cover the outbreak of disasters that had plagued the earth over the last couple of months. Was it her imagination, or had the earth really suffered an unprecedented number of natural disasters lately? Were any of these events connected to each other or the global warming that had been occurring since the latter half of the last century? She made a note to plot all natural disasters over the last two hundred years and compare them when this crisis was over. Or was it only that there were more people on the planet to feel these natural disasters and better equipment to monitor them? Anyway she would find out later; a voice came over the speaker requesting her presence back at the control room.

"What's up?" she asked, as she opened the door.

"They think they have found the plane, but no word on any survivors yet," Markus responded.

"Where did they spot it?" Kristen asked.

"About ten kilometers from where they were last spotted."

"Which direction?"

"South, right along the river."

"That's a bad place, the river widens there quite a bit. They most likely went down in the water. And with the river raging like it is—well, let's hope they're all right."

"We should know soon, they're right overhead. Wait … picking up a transmission now," Markus said, relaying developments. "It's a positive identification, the plane's rudder has been spotted; now the rest of the plane has been seen up the river fifty meters. Apparently the plane broke apart, and the fuselage floated downriver and got caught in a grove of trees. It looks mostly underwater. Transmission is breaking up a little … wait … one survivor for sure. They see her clinging to a tree and waving—now she's pointing, maybe to other survivors. They're trying to get a rope down to her. They've got her! She's being hoisted up now."

Everyone in the control room took a deep breath as the drama seemed to unfold in slow motion. Markus continued, "They're pulling her into the helicopter. We should have a report in just a moment. Here it comes … The plane went down and broke apart. There were six crewmembers on board, and she thinks at least three never got out of the plane before it went under; one other person is alive, and the last crewmember she never saw. The one crewmember alive is downstream, on the bank, up a tree. The chopper is heading there already. They see him … The rope is down—and they've got him back in the helicopter. He says that he saw three of the crew go under with the plane, and they never came up. He saw the other crewmember swimming toward a tree in the river, but he didn't see if the crewmember made it to safety or not."

"Find out who was on the plane and where their next of kin live. We'll have to notify them as soon as we have a positive ID on who is missing, at least for now," Kristen stated.

"I'll get right on it," Natalia replied.

"Thanks," Kristen said. This was the hardest part of her job. She would call their relatives herself. The Squad had very few casualties each year, but the work was dangerous at times, and everyone was aware of the risk. Last year had been a very good year for the Squad with no losses for the entire year. But this year already had claimed the lives of nine members of the Squad before this plane accident.

Since the bodies of the crewmembers had not been accounted for, they were still officially missing. If they were gone, that would be twelve with still over three months to go for the year. Five had been lost in another plane accident rescuing people stranded in a flood in early April in Peru. It was near Cuzco, south of Machu Picchu. The plane went down in heavy fog trying to reach flood victims. The plane's radar failed, and before they could get the plane out of the fog, they crashed into a mountain, and all were lost.

Two others were lost when a building collapsed in an aftershock from an earthquake. There were eight children trapped in the building and before the Squad members were caught in the rubble, they did get seven of the children out safely. The eighth child was lost when the building crumbled in a 5.4 aftershock.

Another member of the Squad was killed when her helicopter went down in a tornado in the central United States. She was attempting to save a family in the path of the storm and would have made it, but the storm increased its forward velocity and caught the helicopter just as it was about to land.

The last Squad member to perish in the line of duty this year had been caught in an avalanche in the Alps after a plane went down on a remote Swiss mountain peak. The member was skiing in to the plane when the snow gave way and swept him down the mountain. He was wearing a radio transmitter, but by the time he was found, it was too late. It was already a brutal year for the Squad, and it was beginning to look like it wasn't over yet. Right then the transmission began again from the subcontinent of Asia.

"We've found another crewmember alive in a tree completely surrounded by the raging river," Markus reported. "He seems okay but is obviously shaken up. He doesn't know about the other members of the crew. He's saying that all he saw was water, and by the time he got out of the plane, he couldn't see anyone in the blinding rain." So at least three Squad members were alive, but it didn't look too good for the other three, thought Kristen, listening to the report coming in.

"Can they land near the wreckage?"

"They're attempting to land now," he relayed to the Squad members in the control room. "They're down ... They're making their way toward the fuselage. They're on the bank near the plane now, but it's hard to approach it with the water raging. One of the members is in scuba gear, and with a rope attached to her she's going into the water. She's near the plane now; she's going underwater near the plane now. She'll be out of communication for a while. We'll just have to wait for a bit until she resurfaces to find out anything."

Meanwhile the other Squad members monitored other parts of the affected area and also checked in on the Atlantic situation.

"Director, please come over here," Natalia asked while she monitored the Atlantic hurricane.

"What's up?"

"Colin just set a record low barometric pressure, surpassing the old record by three millibars. And the winds are reaching record velocity. The Bahamas are really beginning to feel the brunt of this storm. We better send an additional advisory to every possible landfall of the storm now."

"I agree. Also we need to establish a control center near the path. What do you think from the projections now? Orlando?"

"Yes, that would make the most sense for now. It's likely to get pretty bad in Orlando, but unless the storm turns further north, they'll be safe enough to carry out their job. Disney World has offered to give them access to one of their computer control rooms. We can reprogram some of them to the Squad's needs."

"That sound's good," Kristen said. "Send our appreciation to Disney, and let's get set up in the next three hours. Who's the senior Squad member in that region?"

Natalia responded, "It's Alexander Darby. Stationed in Atlanta now, he's already in Florida. He went there when we started moving Squad members into the area a few days ago."

"I know Alex, we served together when I was stationed in Africa in my first year out of the academy. Can you locate him please and get him on the air? We'll need to brief him on the situation and what the plan is going to be."

"I found him. He's in south Florida now at a temporary control center helping with the evacuation going on there. Here he comes on line."

"Alex, do you read me? Over."

"Kristen? Hey, good to see you again, it's been a while; and yes, I hear you loud and clear, and your picture is coming in quite good. What's up?"

"Alex, nice to see you again, too bad it's under such circumstances. But down to business: you are the senior Squad member in the region and therefore are in charge from now on of the Atlantic storm situation. I want you to go to Orlando, to Disney World, and set up a command post for the duration of this thing. Disney is giving the Squad use of one of their computer control rooms and will assist in any conversions, hookups, et cetera, that we'll need to monitor Colin as it makes landfall. We estimate that it's now only twenty-four hours out.

"Also, you should know the pressure is now officially the lowest ever recorded, and winds are reaching record levels also. This is going to get really bad. We need to complete evacuation of all coastal regions and low-lying areas and get all small craft out of there ASAP. The outer Bahamas are getting pounded right now. They were fairly well prepared, and most areas were vacant. Building damage reports are scarce, but what is trickling in doesn't sound good. There appears to be widespread destruction, but fortunately no reports of lives lost to date. Any questions?"

"Not for the moment. I'm on my way to Orlando now, and I'll contact you when I get there. Estimated time of arrival at Disney in ninety minutes. By the way, how's it going in Asia?"

"It's bad. Flooding is the main concern now, and a dam broke several hours ago and has already taken out entire villages. It's progressing downriver now. We tried to warn the people, but we don't know yet if we were successful."

"Let's hope we were. My chopper is here, and I'll call you back in two hours. Out."

"Talk to you in a bit," Kristen responded. She went back to Markus at the Asian monitor to see what was happening there.

"Any news yet on the other three Squad members?"

"Not yet. The diver is still underwater. We assume she's exploring the fuselage to try and find any bodies. Wait—we're getting a report now. She found two dead in the plane, but the third person is not there. That could be good news or bad news. Maybe they got out of the plane and maybe to safety. Let's hope so."

"That's for sure. Can you raise Helene? Let's see how it's going there."

"She's coming in now."

"Helene, how are you holding up?"

"We're fine here, it's the plane that I'm worried about. You've been keeping an eye on the situation, I trust. So you know the bad news, we've lost at least two Squad members and maybe three. This is the first time I've lost members under my watch. I know it goes with the territory, but I feel bad"

"They volunteered for the Squad," Kristen observed. "We all know the risks when we join; and yes, it's hard to take, but let's hope they saved some lives before they went down."

"We know they did. We've gotten reports that the message was received, and many did make it to higher ground before the wall of water reached their village. So their sacrifice saved lives. Have you notified their families yet?"

"Not yet, we wanted to get a positive identification before we called their families. The diver brought up the dog tags of the two bodies she found, and we have their names now. I'll go call their families now."

With that Kristen left the control room and went to her office to do the very worst part of her job. The calls were brief and to the point, with as much empathy as she could convey over the phone. She would now contact personnel for the Squad, and they would go over in person and help with the funeral and memorial arrangements. Silverton, sensing her grief, jumped on her desk and began rubbing against her hands, which were tightly clasped. The calico starting purring, and Kristen petted her on the top of her head.

"Thanks, my friend, you always seem to know when I need some comforting of my own," she said to her cat. With that done, she returned to the control room with her little friend right behind her. She was quiet but knew she would have to pull it together and be the leader, energy, focus, and whatever else it might take to do her job and keep the Squad on their mission.

"Director, I have a new report on Colin," Natalia said, calling Kristen over.

"What do you have?"

"The pressure is now eight millibars below the previous record low of Typhoon Tip in the Northwestern Pacific in 1979," Natalia related, "which was 870 millibars or 25.69 inches of mercury, and the winds have reached a new official record for a tropical storm surpassing the old record of Typhoon Nancy in 1961 of 215 mph (346 kph). This storm is going to cause major damage if it goes through the Miami metropolitan area. We've sent all this information to Alex already and every other government agency that needs to know what's going on with this monster. We really need to get every living thing out of south Florida, and we need to do it now."

"Agreed. We have to step up the evacuations right away. Can you get me the governor of Florida?"

"Yes, stand by, it will take a moment. Here she is."

"Governor, this is Kristen Verba, director of the Global Disaster Squad. We have been monitoring Hurricane Colin in the Atlantic, and you should know that the storm has reached record strength in both pressure and wind speed. This has the potential to be the worst storm of all time bearing down on your state. We would suggest mandatory evacuation now of all inhabitants south of West Palm Beach and recommended evacuations of all people south of Orlando. We don't know for sure where this storm is going to come onshore, but on its present path it's going right through Miami. It could of course veer north or south and affect a possible wide range of territory, so you need to have the entire state on alert. Also there likely will be huge tidal surges, so all coastal areas need to be empty within twelve hours, and there is a high probability of tornadoes in the next two days. You should try to get as many people as possible to the interior of the state and in heavy-duty shelters wherever possible."

"Thanks for the data," the governor said. "I figured as much, as I've been watching the reports coming in from the National Hurricane Center. We've moved almost everyone out of the coastal regions already and are trying to get the low-lying areas cleared next. We are trying to avoid a general panic, which might clog all the main highways out of southern Florida. We're trying to keep the flow as close to maximum as possible without creating gridlock. So far it's going well. We're going to need water shipped in right after the storm passes, as it's fairly likely that most water sources will be contaminated. Also most grocery stores have had a run on them, which has depleted their stock to near zero, and no trucks have entered the region in the last twelve hours so we may need food flown in as soon as the storm passes. Anything else, we'll call you." The governor was in her first term and was facing the most formable hurricane since hurricane Andrew in the early 1990s.

"Sounds like you have the situation under control as much as possible, given the circumstances," Kristen said, feeling more confident because the governor had already made a lot of positive moves. She continued, "The Squad has set up local headquarters at Disney World and has converted several of their computer command centers to the needs of the Squad. I'll give you their numbers, as they're ready to assist you in any manner they can. They'll coordinate efforts with you to get as many people as possible to safety. Anything else we can do for you, just let us know. We'll keep in touch. Good luck," Kristen said signing off to the governor of Florida.

"Thanks, Director, and we just want you to know how much we appreciate all that the Squad does for everyone. Talk to you later."

"Helene's coming on the air; she's calling for you, Director."

"I'm on my way. Helene, how's it going?"

"We're very excited, and I thought you would want to know, we found one of the missing Squad members, and he is alive with only a broken leg and some bruises. He was thrown from the plane before it broke apart and sank. He managed to make it to shore, and we found him before the water level rose from the broken dam, which it did only minutes after we found him.

"By the way many people did make it to safety before the water passed, but I'm afraid that many others didn't. The water took out all of the buildings in at least eight villages along the river. The flood surge is subsiding now as the river widens, although it still is doing some major damage along the banks; but the really good news is that the rain is abating along with the winds, at least there along the coastal sectors."

"That's great that you found one of the missing Squad members. Send me the details, and I'll relay the information to his immediate family. How's the weather where you are?"

"It's still pretty intense here. Winds are over two hundred kilometers an hour, and the rain is still coming down at over four centimeters per hour. The rivers are all beginning to reach flood levels, and I'm afraid that the aftermath of this rain will be rather intense. We're likely to see flood stage exceeded by as much as twenty to thirty meters in some areas. That will result in even more evacuations needed and severe damage along the floodplains. We'll get going on the extra evacuations as soon as the weather allows it. It'll be a while yet until that happens, so I've got the majority of the Squad members here on break; most are sleeping, getting some much needed rest to be ready for when the storm breaks."

"That sounds good. By the way, have you gotten any rest yourself yet?"

"No, but I'm going to right now, at least for a few hours. I'll check back in after I get some shut-eye. Talk to you later. Call me if anything changes. Out for now."

"Talk to you later."

"Director, you better see this," Natalia called over to Kristen after disconnecting with Helene.

"What's up?"

"The Atlantic storm is veering more toward the north. It looks like Orlando is the now in the eye's path. The Bahamas are in for a more direct hit also. I've sent word out to everyone affected. Nassau is directly in the main path. Winds are beginning to pick up there already, and tidal surges are reaching three meters as we speak. Coastal flooding has begun. The place is going to be very hard hit. Winds are well above record levels for any tropical storm ever recorded, and the scary thing is that the pressure is still falling. This is going to get even worse"

"Get Alex on the screen, please. We need to get going on evacuating central Florida and all coastal regions as far north as South Carolina and as far west as Houston just to be on the safe side. There won't be much time when the storm finally does make landfall to get people out of danger."

"Here's Alex now."

"Director, we just got the word on the storm's change of direction, and we've got everybody on it now," Alex said. "But this is putting pressure on the entire infrastructure; every highway, railroad, and airport is full to capacity already with the people leaving from southern Florida. Plus we've got almost every road one way headed north to allow as many people as possible to get out of dangerous areas. We've had to restrict southbound traffic to emergency and official business only. How long do you project we have until landfall, given current data?"

"Less than twenty-one hours and maybe as little as eighteen, as it seems the storm is increasing forward velocity ever so slightly."

"Doesn't give us much time, but we should be able to get quite a few out of here. Road speeds on the interstates and the Florida Turnpike are down to less than sixty kilometers per hour. We were hoping to keep the speeds up around ninety to move as many people as possible, but the roads are just too backlogged to get them going any faster."

"What about small boats in the storm's path?"

"We've had restrictions on small craft for days now as the seas have been rough for that long from the swells being pushed ahead of the storm, so we're in pretty good shape there. Most of the larger ships like oil tankers, cruise liners, and cargo ships pulled out ahead of schedule and are well on their way to safe waters. We've brought as many supplies as we could get in already for after the storm's passage like bottled water, food, medical supplies, et cetera, in case the infrastructure is badly damaged. We will probably need more flown after the passage of the storm. We'll be as ready as we can be, but given the record strength of this storm, that may not be enough to save building structures, forests, farm lands and so forth.

"The national and state parks will likely be hit hard, and many endangered animals and plants will suffer. Zoos and animal shelters have been evacuated wherever possible. We urged people to take their pets with them when they left their homes so we don't have pets loose after the storm.

"Hospitals are fully stocked with emergency supplies, fuel for generators, extra blood, and so forth. Most newer public schools, which have been built under the new codes that require them to be able to act as shelters in emergencies, are fully staffed and stocked with enough supplies to last for a couple of weeks. So you can see we're pretty much ready. Now it's just down to how many people we can get out of the storm's path. Those who can't leave will be moved to one of the schools we have ready. Anything else, Director?"

"Not at this time. It looks like you've got the situation under control. How many Squad members are in the region?"

"We have two thousand within the state of Florida and another seven thousand in all the areas that might be affected from the Carolinas to Texas. Then there are a few thousand more in the rest of North America who could be brought here if we need them, but for now they need to stay stationed in case of other emergencies and those fires that are still burning in the West."

"Well, that sounds like plenty of help. We'll start checking in with you every hour on the hour, so unless something changes dramatically, talk to you later. Out for now."

"Sounds good; bye till later."

"Natalia let me know—"

"The minute anything changes with the storm," Natalia finished for the director, knowing what she would be requesting. "Yes, ma'am," she added.

"Thanks, you know me well," Kristen said and turned to check in with Helene, who was still in rest mode. Kristen decided to let her rest, as there wasn't much she could do anyway until the storm let up a little more.

"Well, we'll just have to wait for nature to play out its hand now and respond accordingly," Kristen said. "We've done all we can. I'll be in my office for a bit and will be back in time to talk to Alex. A few of you should take a rest break and be back here in four hours when it really gets going."

All but three of the staff went to take a break, leaving one to watch the Atlantic storm, one for the Asian storm, and one covering the rest of the world. They didn't realize it at the time, but this would be their last rest break until the whole story was played out.

CHAPTER TWELVE

It was now nearly two weeks since the first sighting of the little tropical wave off the coast of Africa that would later become Hurricane Colin. Landfall was expected soon as the storm crept toward the mainland of North America. Kristen Verba napped for thirty minutes on the small cot she had set up in her office with Silverton sleeping with her. She had lain down contemplating the last two weeks and finally in utter exhaustion drifted off. She didn't dream. She awoke slowly, forgetting for a moment where she was. It didn't last long as the speaker in her room summoned her to the control room.

"Director, it's Helene, she's got an update on the Asian situation."

"Thanks, go ahead Helene."

"The rains have let up on the southern edges enough for us to get out and start evacuating civilians along rivers likely to be flooded very soon. That's the good news. The bad news is that this seems to be a temporary break in the storm with a strong squall line approaching in three to four hours. This doesn't give us much time, and we'll have to concentrate on putting the word out to the population and letting them get to higher ground on their own. We have planes and helicopters in the air already and are in radio communication with some of the officials in the areas under threat. The winds have subsided here for the time being, but they will probably pick up when the squall line swings through. We'll have to have the planes back on the ground by then or face problems like we had earlier.

"The first reports of damage and causalities are coming in as radio communication is slowly being patched up. Building damage has been severe, especially along the Indian coast south of Calcutta and through all of Bangladesh. Some areas were wiped completely clean of all structures, but loss of life there was minimum due to the early evacuations along the coast. However, in other areas there have been great losses. Mostly from the flooding along rivers and in particular the river where the dam broke upstream. There are reports from there of as many as two thousand dead and thousands more missing. Other areas are reporting nearly as many casualties, almost all due to flooding along rivers.

"There were other losses with tornadoes, a bus that went into a river when the bridge it was on collapsed while it was passing over, killing more than a hundred people, and of course our own losses in the Squad. This has been bad, but easily could have been worse. We could have lost millions if we hadn't evacuated as many as we did. I know that's no consolation to those who did perish and their families, but—"

"We've done the best we could have considering the situation," Kristen put in. But there is still a lot to do, so we'd better get busy. Good luck, and stay in touch."

"Thanks, I'll talk to you soon. By the way how's the Atlantic situation?"

"It's getting really scary. The pressure keeps setting new record lows, and the winds have now reached record speeds too. The outer Bahamas are already taking a beating, and Nassau is directly in the path. From there it looks like central Florida, but we don't know for sure as the storm keeps switching course almost every hour. This means we have to evacuate most of Florida. It could be bad, really bad. We're ready, at least as ready as we can be. There's only so much we can do, and the rest is up to Mother Nature. I'm about to check in on the Squad leader there now. Talk to you soon."

"Who's in charge there?"

"Alex Darby. Do you know him?"

"Yes I do. Tell him hello, and wish him luck. Sometimes that's what it comes down to with Mother Nature."

"I'll tell him that. Talk to you later. Update on Colin, please," Kristen asked Natalia as she approached the Atlantic monitors.

"It's still heading for central Florida through the Bahamas, although it's turned a bit more southerly at the moment. We're pretty confident at this time that it won't turn north, as high pressure is building off the coast of the eastern United States, which will block any movement that way. This makes it more likely that the storm will cross Florida and make a second landfall somewhere in the Gulf of Mexico.

"Also winds have reached hurricane levels in the inner Bahamas," Natalia continued, "and tidal surges are at ten meters in some areas. Surges are reaching northeastern Cuba now, and gale-force winds extend out from the eye two hundred kilometers in all directions. Even winds in southern Florida are picking up velocity.

"Another thirty to forty waterspouts have been detected, but none made landfall. We're bracing for a lot of tornadoes when this storm reaches land. Already the total of tornadoes and waterspouts is close to a hundred, counting the ones that hit the Dominican Republic."

"Estimated landfall time in Florida?"

"At the present speed and direction ... let me see—fifteen hours, plus or minus an hour. It has increased forward velocity and is aiming just north of West Palm Beach at the moment, which is a closer landfall than further up the coast. We have a hurricane warning out from Key West to Jacksonville and a watch out from Cuba to Georgia."

"That sounds good for the time being. Looks like it's time to check in with Alex." As soon as he came on the screen, Kristen continued, "Alex, how's it going?"

"We're doing all right. All effort at the moment is on getting people out of southern and central Florida. We're thinking we've got only about eight hours before we close all roads and get everyone into a shelter. That will give us about five to seven hours before landfall, as the winds and rain will likely be fairly intense by then.

"I know the Bahamas are being pounded. The last report I heard, about twenty minutes ago, had the eye of the storm just beginning to come into the eastern Bahamas with winds above 285 kilometers per hour. Tides were as high as fifteen meters, and most communication lines are out. We had a radio link up to about twenty minutes ago, but even that connection broke. At the moment we don't have any ties to anyone in the Bahamas. We've got about a hundred Squad members over there, and I'm sure that as soon as they can, they'll be in touch with us. Every coastal area was completely evacuated, so although we are expecting extreme property damage, we're not expecting much loss of life. Let's hope that remains the case."

"That's for sure. The Bahamas have been preparing for this sort of storm over the last ten years, ever since that last really bad storm. I've got to go and check on final preparations. I'll call back in an hour. If the storm gets too intense here earlier, we'll go ahead and close the roads and shut this place down. We'll be playing this by ear from now on. Talk to you later."

"Sounds good. By the way, Helene says hello and good luck. She's in charge of the Asian Squad effort."

"Tell her thanks, and the best to her too. Bye for now."

"Bye."

"Director, Helene's back on the line."

"Helene, do you copy?"

"Yes I do. We've had a major dam break in central northern India. And the rain is still coming down at three to four centimeters per hour. The rivers are all at flood stage, and this dam that broke flows into a tributary of the Ganges River. The water will be to the Ganges in about three hours. We had evacuated most areas along the Ganges anticipating that it would flood, but the tributary is another story. We've sent word to villages downstream of the break, but I'm afraid they won't have time. The water is moving quite fast, and communication in the area is limited at best. The water will be to the

Ganges before we could have a plane there, so about the only thing we can do from here is radio warnings to low-lying areas to get out of there now. We've also sent word to areas downstream of other dams that there could be problems and to move to higher ground at their earliest chance. There are some good-size villages in the path of this wall of water, and we can only hope they had time to clear out."

"Have you received any communication back from anyone in that area that they've heard the warning?"

"None at this time. So we don't know what's happening. Our closest Squad members have responded that they picked up our message, but they're about a hundred kilometers from the river and don't have any firsthand knowledge of what's happening there."

"Well, keep sending out the warnings," Kristen said. "I don't see much else we can do for the moment. We'll need to get rescue people in the area as fast as possible, but that still could be hours from now by the way the storm is progressing. Our satellite images show it slowing as it moves inland, but still maintaining its circular rotation, which is pulling moisture up from the Bay of Bengal. This makes continuing rain and wind all but certain. We were hoping that as the storm hit land it would begin to break up and weaken, but that hasn't happened yet. It still is a class-five hurricane."

"Yeah, we're still being pounded here. It did slow up for about an hour, but then came back as strong as ever. We'll stay in touch; I'm going to check on our message attempts. I'll call you if anything else happens or changes."

"Roger that," Kristen said, signing off to Asia. This was really getting bad. The loss of life had been kept to a minimum until now, but with this new dam breaking, that was likely to change, and change in a hurry.

"What's the size of the villages downstream from where that dam broke?" she asked Markus, who was monitoring the Asian storm.

"Here's the data now. Looks like there are about twenty-six distinct ones, and there seem to be a lot of small communities along the entire river. Total estimated population in the area is … I'm adding it up here. Let's see, around four hundred thousand people." Markus went silent for a moment as his words sank in. There could be hundreds of thousands of casualties now from this storm, making it the worst natural disaster of this century and one of the worst of all time and it wasn't over yet.

Kristen said, "We'd better put out a call to all reservists, standbys, and any retired Squad members who are available to come on to temporary active duty throughout the world. It looks like we're going to need all the help that we can get."

"General call Number One activated worldwide ... now," Markus reported as he activated a special code for just this situation.

"Good. Now let's start planning where they need to go." Kristen took a deep breath. "Everyone in the Eastern Hemisphere needs to come to India to help with the aftermath of Nisha. We'll need to get large groups along the rivers where the dams broke upstream to start rescue and reconstruction of basic infrastructure. Other groups will need to move into coastal areas and assist with the relocation of local populations and supply distribution. Medical teams will have to move throughout the region and help where needed."

Markus replied, "We can set up three basic areas for a coordinated search and rescue sweep of the affected regions. North, east, and west should be the easiest designations considering the lay of the land around the Bay."

"Good thinking," Kristen said. "Let's put 40 percent of the Squad members in the north section and 30 percent each in the west and east zones. The worse damage is in the north, which is why we'll need the largest number of members there. Find out who the three senior Squad members are, after Helene of course, and we'll have each one of them head up one of the areas. Then we'll have them coordinate their efforts with Helene and all local officials to maximize the rescue effort."

"Here are the names now, and they're being contacted to move into position. I'll feed the names and plan to Helene and have them contact her as soon as they're in place. Anything else I should do?" Markus asked.

"Not at the moment; we'll wait until everyone gets in place. Meanwhile let's go ahead and get North America divided into similar zones except let's have four zones there: Florida, west of Florida along the coast to Texas, south of Florida including the Caribbean and the Bahamas, and the fourth area should be the inland United States wherever the storm has the greatest chance of making second landfall. Find the four senior Squad members, have them report in to Alex immediately and start planning for after the storm's passage."

"I'm on it."

"Great. Now the rest of us need to plot out every contingency for all the scenarios that we can develop for both storm fronts and be as ready as we can be. These are not normal situations, if anything can be construed as normal anymore; so we need to alter our thinking and responses to these storms."

"You can say that again. We've got data now that shows that right before landfall Nisha reached the second lowest barometric pressure ever recorded, second only to Colin. That makes these two storms the strongest ever recorded. And here we have the two record storms at the same time. Wow!"

"Then we'd better get busy," Kristen said. The next few hours were spent making final preparations for the Atlantic storm and shifting newly arrived Squad members into position in Asia and North America. There were a lot of logistical considerations to take into account. Squad members would need to fly in supplies, procure rebuilding equipment, and then get ready for the unpleasant tasks of body recovery, identification of the casualties, and disposal of the victims. Then there was the long-range planning to restore the affected regions to something resembling normalcy. Trees would need to be planted, crops replanted, beaches rebuilt, wildlife habitats revitalized, dams reconstructed, rivers returned to their previous course and within their banks, homes erected, and more. It would take years to remove most of the visible signs of these storms, at least the Asian one.

The storm in the Atlantic had yet to cause much damage, but the potential for devastation was high, what with wind speeds well over three hundred kilometers per hour and the sheer number of tornadoes being spawned by the storm. The next three to four days could be catastrophic for the world.

The director took a few minutes to make sure that all pressing paperwork was current and worked ahead a few extra days in case she couldn't get back to her desk for a while. While she worked, her constant companion was lying on the desk, cleaning herself and all the while purring.

Markus called her there in the office. "Director, Helene's back on the air and wants to talk to you."

"I'm on my way," she said, setting her papers in order, and headed back to the control room.

"Helene, this is Kristen. What is the latest situation?"

"I wanted to update you on the dam breaking. It has been real bad. Very few people had a chance to get to safety higher up the river, and first reports coming in show almost everyone was swept away by the floodwaters. Further downstream the survival rate was higher, but still at least half the population was lost according to witnesses in the area.

"We have further accounts that most structures have been destroyed, every bridge crossing the tributary was taken out, and in several areas the stream has changed its path. The flood wall has weakened somewhat as some of the water has spilled out into surrounding areas, and the channel is being dug out wider also."

"How long until the water reaches the Ganges?"

"It's about an hour away right now. We've had better luck getting people out of the flood plain of the Ganges, as many had already anticipated flooding and moved to higher ground. Plus we have had more time to get out further communications to those endangered areas."

"How many losses are you estimating at the moment?"

"Somewhere between two and three hundred thousand may have been swept away. It will take days to get any closer number, and the truth is we'll probably never know for sure as all the records are gone as well. And where everyone in a village was wiped out there won't be anyone to notice if anyone is missing. So it's going to be hard to get a real count."

Kristen was stunned when she heard the report. The magnitude of the this disaster was sinking in.

Helene went on, "One good report we can give is that the storm has been lessening somewhat for the last hour."

"That is good news. The satellite images confirm that the storm is lessening, at least the wind velocities. The rain seems to be just as strong as ever. It will probably take a while for the precipitation to let up, which meantime will just add to the flooding. When do you expect to get airborne?"

"We should be up in maybe another hour or so if the wind continues to abate. We're ready at a moment's notice. Wait, we're getting new information from Bangladesh. They're reporting tornadoes, six, maybe eight, moving toward Dhaka. They're big … up to one hundred meters across."

"How far out from the city are the twisters?"

"About twenty-five kilometers?"

"And how fast are they moving?"

"Can't tell for sure, but around forty to fifty kilometers per hour, so they could be to the city in thirty or forty-five minutes. We've sent word to everyone in the region, but it's doubtful that many people are in communication with the outside world. Hopefully some will hear and spread the word."

"Keep us posted. We're getting news from Florida, so I'll be back in a bit. Out for now."

"Director, Alex is coming on, here he is."

"Alex, how's it going?"

"We're already picking up some pretty strong squalls. Four more tornadoes have touched down here and there, but no severe damage has been reported. And we've started the shutdown of all highways in southern Florida and will in central Florida soon. The show is about to begin.

"We've heard reports from the Bahamas of incredible damage. In some places every structure has been leveled. On one of the smaller islands south of Nassau every tree has been knocked down, and only one building is still standing. It sounds pretty bad in the Bahamas in terms of structural damage, but so far, the loss of life has been minimal. A lot of people were moved off the little islands and took shelter on the larger islands, which are better suited to take this kind of pounding. I'm sure we'll have some casualties though, simply due to the raw strength of this storm. We're trying to get as many people here in Florida into tornado-resistant shelters now as we can, to prepare for the mess coming toward us.

"Here's a report from southern Florida of the first gust to reach hurricane strength. We're now predicting that the storm's eye will pass near or through West Palm Beach based on its current path. That would spare the most populous areas of Miami and Fort Lauderdale, but there are still a lot of people in its path. Plus Tampa and St. Pete on the Gulf side are both likely to be hit hard on this present course.

"More reports coming in from southern Florida," Alex said glancing at the computer screen in front of him. He continued, "Winds are definitely picking up—sustained winds now approaching hurricane strength, and the rain has began in earnest. I think we'd better go ahead and close all highways now throughout central Florida as far north as Gainesville and get people into shelters.

"Well, Director, wish us luck, we're going to need it. We'll try and keep in touch during the storm, and we've placed a lot of remote cameras around the state to monitor the situation as it develops. They'll be on a revolving viewing schedule starting clockwise from Jacksonville down the coast to Key West and back north through Tampa to Panama City, with detours into the interior of the state as the rotation proceeds. This should give us a pretty good picture of what's happening as the storm passes through. This footage will go the Squad's research library for complete analysis. Well, talk to you later."

"Sounds good, and good luck." Kristen signed off with Alex and went back to Markus, who was monitoring the Asian situation. "Heard anything new?"

"Yes, those tornadoes are heading straight into Dhaka in about five minutes. Helene had the word sent to take shelter, but we're not sure how many people heard the warning. Wait … we've established contact with someone in the path who has an old ham radio. The message was heard in scattered areas, and word of mouth spread it further, but there wasn't much time, and the heavy rains are still hampering travel.

"They can see the wall of tornadoes approaching now. It sounds like they've gotten bigger. The eyewitness, a woman in Dhaka and not a Squad member, is reporting that the twisters are almost in a straight line, ripping everything in their path to pieces. She can't make out for sure how many individual tornadoes there are, but she is guessing eleven from how the damage is progressing. She has a view from a building off to one side and is watching through binoculars. The line is moving through the northwest section of the metropolitan area, which is fairly densely populated, but so is most of Dhaka. The destruction is intense, she says. She reports seeing buildings explode into the air, the sky full of debris whirling around at extreme speeds. It sounds real bad."

"Who's in charge in Dhaka?"

"Jireco Chirano. He's a Bangladeshi national and knows the country very well."

"Can you get through to him?"

"I'll give it a try. He's not in the control center at the moment; he's on his way to the damaged area already. We've got a page out to him now. He should be receiving our call any minute."

"Good, patch him through as soon as you make contact."

"We've just made the connection. Here he is now."

"Mr. Chirano, this is Director Verba at Global headquarters. We've been tracking the line of tornadoes. What can you tell us?"

"I have been following the storms by boat—that is the only way to get around right now due to all the water—and they're moving through one of the most populated sectors of the city. There was no way to get the people out of the way as the line is so wide, maybe as wide as ten kilometers; it's difficult to tell from our perspective as we're following the storm line through the city. We tried to get people in basements, low-lying areas, or into the center of buildings wherever possible, but as you may know, Dhaka has a high water table, so there aren't many subterranian structures in the city or low-lying areas for that matter.

"I'm seeing widespread devastation, people running through the debris looking for family, friends, neighbors. The number of bodies lying scattered about is staggering. I can see several hundred bodies from where we are now and maybe a thousand so far since we entered the tornado destruction area. At this rate we may be looking at a hundred thousand or more lost. This is a terrible blow to this city, which had been pretty beat up by the typhoon and is getting flooded out by all the rain-swollen rivers moving through."

"What can we do to assist you?"

"Well, I know the Squad is stretched to the limit right now. I've been following the worldwide call-up, but obviously we need help here as soon as we can get it. Looks like medical first, then personnel and equipment for rescue operations, food and temporary shelters, followed by rebuilding assistance. It will take quite a while just to find all the bodies, and then we will have deal with them. Buildings are lying completely in heaps, which will all have to be sifted through for survivors."

"I'll shift some of the Squad members from less urgent projects to your location to assist with the rescue efforts," Kristen said, pulling up a list of Squad member deployments. She continued, "We'll scan the Eastern Hemisphere and see where we can pull some help. We've got some in Thailand replanting from that tidal wave a while back who are on their way; the fire in Australia is under control and those Squad members are on their way. Let's see where else we can find help … Oh yes, here are some in the China earthquake relief effort that could come there temporarily to assist in the rescue. That should do it for now; that's close to a thousand personnel and hopefully will be adequate to find all the people missing."

"Let's hope so. When do you think they'll reach here?"

"Within four hours the first will be in place and the rest will be there in a few hours more."

"That'll be great, thank everyone for their help. I'll check in a little later."

"Okay, we'll be in touch," Kristen said and moved back to the Atlantic monitor. "Where is the storm now?" she asked Natalia, who was coordinating the information on Colin.

"The eye of the storm is in the middle of the Bahamas right now, on course for West Palm Beach. Florida has shut down completely; we overheard the call some fifteen minutes ago for everyone to get off the roads and seek immediate shelter. All highways, roads, trains, et cetera, are closed to any further traffic. The cameras have been turned on, and as you can see it's a ghost town everywhere across the state except in the very northwestern panhandle where they still have a little time before being affected by the hurricane. We have Alex on the monitor now, Director."

"Alex, I can imagine you're as ready as you can be. What's the situation from there?"

"We're ready; the waiting is the hardest part. The sky has gotten that pre-storm eeriness—you know what I mean, the sky has gotten quite dark, and for the moment the wind is barely blowing here, although along the coast we have reports of winds already gale force and a few gusts to hurricane strength. The rains are picking up in advance of Colin as well as the tidal surges. We anticipate the storm will start in earnest here at any moment. The wind just kicked it up a few notches as we're talking. Can you pick up the cameras around the state?"

"Yes, we've already been watching them as they rotate around the state. Looks like ghost towns from an old western movie."

"Yeah, that's what we thought too."

"Well, good luck to all of you, and we'll contact you again soon. If anything drastic happens, you can reach us here immediately; otherwise we'll be in touch on the half hours until the storm has passed through Florida. Bye."

"Sounds good. Over and out."

Kristen left the control room for a routine check of the incoming messages that went to her office directly. These were mostly administrative communiqués from various nations, either requesting money or succor for a particular project under way within their borders or in rare instances actually offering money or service back to the Squad. Then there were the thank-you notes from all the people, cities, states, nations, companies, villages, and parishes that the Squad had helped in recent times; sometimes there was even money from one of those grateful entities.

Checking through her phone messages today, however, brought her to tears as she heard her mother say that her uncle had passed away from a heart attack earlier in the day. Kristen had always been close to Uncle James, as he had been one of the people besides her parents to encourage her to join the Squad after college and to further seek it as a career. Now he was gone, and she wouldn't be able to attend his funeral because of the crises.

He would have understood. This was her finest moment. She was overseeing the largest number of active-duty Squad members ever mobilized at the same time, orchestrating the largest relief effort ever attempted by the Squad; she was facing the two worst storms in the recorded history of the planet, in addition all the earthquakes, floods, breached dams, forest fires, tsunamis—name a disaster, and it had happened this year somewhere on the globe. It had been a taxing year for the earth, and she had guided the Squad through it with comparable ease. She was one of the best directors the Squad had ever had, but at this very moment she wished she were home in Australia with her family. Her moment of solitude was quickly broken with a call from Natalia in the control room asking her to come as soon as possible.

"I'm on my way," she said and swiftly returned to the control room, but not before leaving a brief phone message to her mother conveying her condolences and her regrets about not being there; she promised to visit them as soon as the circumstances would allow. Then she was in the control room.

Her feline friend never left her that day, always following the director wherever she went through out the building. Silverton only stopped purring when Kristen heard that her uncle had just passed away. She was purring again, as if to say "life must go on."

"What's up?" Kristen asked as she entered the room.

"It's Jireco Chirano," Natalia answered. "He's sending some preliminary information on the number of victims in the region the tornadoes had struck. I'm patching him through to the big screen now."

"Mr. Chirano, how are you? What's your status?"

"We have scanned the area that the twisters passed through, and the first estimate is near three hundred thousand missing and presumed dead. The good news is that the tornadoes have dissipated and are no longer a threat to anything. Our teams here are already in the rescue effort and are pulling bodies from the wreckage as fast as we can. The endeavor is being hampered by the continuing wind and rain from the typhoon, but fortunately that is letting up for now too.

"We have found a few survivors," Chirano continued, "but for the most part the people we are finding are already dead. We are setting up triages for the wounded and morgues for the dead. We will be needing heavy equipment in here to help with the excavation of the collapsed buildings and in the disposal of this number of bodies. Can we get additional front-end loaders, dump trucks, backhoes, and the like in here ASAP?"

"We'll get right on it. Anything else besides medical, food, and water?"

"Yes, with the number of casualties and the high water table in this area, we are suggesting cremating the bodies to avoid contaminating the water supplies and to impede any spread of disease. In order to do that we'll need additional fuel to carry this off. I've talked with the local officials, including the religious leaders, and they've all agreed to this course of action as the best for the survivors. Some of the religious leaders were hesitant at first but were eventually convinced by the magnitude of the situation, which they could see after touring the area where the tornadoes hit."

"We'll see what choices we have and get back to you as soon as possible. Meanwhile we'll send our sympathies to the people of the region," Kristen said, thinking for just an instant of her own recent loss, "and get that heavy equipment there as soon as we locate some we can divert to you. We're hopeful you'll find more survivors."

"We're hoping too. I'd better get back to the job at hand, and we'll be in touch if anything changes. Thanks for your help from all the inhabitants of this part of the world."

"You're welcome, but we're just doing our job. Thank everyone for assisting in this rescue effort. Bye for now." And as he went off another call was coming in.

"Director, Helene is coming in now. Here she is …"

"Kristen, we've been following what's happening in Bangladesh, and we can send some help in from here for the time being. The rain is still falling here, and we can't do much for now until the rain subsides some more anyway, so I've dispatched some of our members to Dhaka. They left here a few hours ago and should be there not too much longer. Our hearts go out to the people there, they've taken a double beating in this storm, and they may be in for a third hit from floodwaters heading their way.

"The storm is beginning to weaken somewhat, at least the wind velocities; but the rain is still coming down pretty hard as the storm tries to rain itself out. I'm afraid that in many areas the worst is still yet to come as the floodwaters begin to peak over the next few days. The storm has pretty much halted any movement and is just pumping moisture up from the Bay of Bengal into the subcontinent. This is a fairly typical storm pattern here, but not at this level of intensity. I just hope it ends soon to avoid a truly major disaster. How's the Atlantic situation?"

"Colin is crossing the area from the Bahamas toward Florida probably just about now. I'm about to call Alex and see what's happening. We're hooked up to cameras around the state, and we already show the wind pretty strong in southern Florida and rising gusts throughout the rest of the state. The winds near the eye have topped the strongest ever recorded, and the pressure is well into record territory for the lowest ever registered. The next twenty-four hours will be hard on Florida as the storm passes through the state. We're showing a computer projection of the winds possibly diminishing as the hurricane passes over land, but it's not likely because Florida is so narrow. We're expecting winds near 350 kilometers per hour as the eye approaches what now appears to be West Palm Beach area. Now we just have to wait it out."

"Wow, that sounds bad," Helene said. "It's hard to believe that both of these storms are occurring simultaneously on either side of the world; or maybe that's why they're happening together. The energy patterns of the weather systems of the world might just be balancing themselves out with these extreme energy releases. It will be interesting to see what we learn from these storms."

"When all is said and done, I don't think any two storms have ever been watched, examined, recorded, and just overall investigated than these two cyclones have been. Your theory on why these are happening together sounds like it's worth checking into a little further. We'll talk about it again when this is all over.

"I'd better check in on Florida now, as I'm scheduled to call Alex about now. I'll call you back in one hour or earlier if something changes. Out for now," Kristen said.

"Give Alex our best; talk to you soon."

"Director, Alex is standing by now," Susan stated as Helene's picture faded.

"Patch him through to the monitor, and put the pictures from the scan cams on the big screens. Alex, how's it going?"

"As you can see from the scan cams, southern Florida is beginning to get pounded, and the rain is falling at three centimeters per hour now with winds well above hurricane strength from Miami to the Cape. Tidal surges are topping five meters; power is off throughout the region already. We have turned power off in many regions as well as gas supplies. We did pretty well getting everyone into some form of shelter before the winds picked up, although we had a little trouble with some people who refused to leave their homes. We convinced a few of them to leave when they understood the true nature of this particular storm. Still a few refused to leave and they're still there. We wished them luck and left.

"Another report is coming in from our West Palm Beach team; they just had their first gust to two hundred kilometers per hour, and sustained winds are at 130 kilometers an hour. Winds everywhere in southern Florida are reaching one hundred kilometers per hour, and rain is falling very hard now in all the same areas. We're tracking several tornadoes also on our triple Doppler radar systems. There are at least three on the screens now moving inland north of West Palm Beach. It doesn't appear that they will hit any major population areas at the moment, but it's a narrow path that they must travel through to avoid some major damage.

New reports coming in from the coast show the winds are steadily increasing. Tides are reaching flood stages already along the eastern seaboard of Florida. Look at camera 32 just north of Fort Lauderdale: the tidal surge has reached across the A1 Highway and is lapping up against those businesses along the west side of the thoroughfare. Even Sarasota on the west coast is starting to show wind velocities approaching eighty kilometers per hour. We estimate another hour and this thing will be out of control. Are you following the scan cams around the state?"

"Yes we are. Some of the cameras are starting to shudder a little in the wind, giving us an idea what's happening there. Where is the southernmost camera again?"

"Key West, at the Hemingway house. They already had a security camera installed and were gracious enough to let us hook up to it to monitor the storm. Here's another report coming in from West Palm ... Tidal surges are reaching well into the city now ... Winds are gusting to three hundred kilometers per hour ... Buildings are starting to show damage ... Those tornadoes veered into a small town and have pretty much leveled half of it ... Several accounts of injuries are coming through, but no deaths at the moment. Street flooding is now widespread throughout the city. Trees are toppling, roofs are being torn off, telephone lines are down everywhere. We're switching to the scan cam there to monitor the situation. Are you picking up the transmission?"

"Yes, and the weather data flowing in from the remote sensors at each camera. It's looking like the scenes we saw from Asia yesterday. The horizontal rain, the trees bending to the ground, and the sound—the deafening sound. Alex, keep on the air, and I'll be back in moment. I need to check on Asia."

"Roger, we'll stay on as long as we can."

"Natalia, let's split the screen between Asia and Florida and follow both on the big picture."

"Okay. Do you want Alex on or a rotation of the scan cams?"

"Let's split the Florida screen between scan cams and Alex at their local control center." Kristen turned and asked Elawa, "Do we have a link with Helene?"

"Yes, she's coming through now." Meanwhile Natalia put the feed from Asia on half the screen, and the other half had four smaller screens: one on Alex and the other three rotating scan cams through Florida.

"Helene, what's the situation?"

"The wind is waning, but the rain is increasing again. We have major flooding throughout the southern region of Indian and all of Bangladesh. Streams are over their banks; rivers are well above flood stage across the board, even lakes and reservoirs are near to overflowing virtually everywhere. We're at the breaking point now, and our resources are taxed to the limit.

"Even if we could get some help in to the stricken areas, where could they get the people to safety? The roads are underwater, the rivers impassable, the airports are out of commission. We've given out warnings continuously, and all we can hope for now is that people are seeking higher ground. It's pretty much in the their own hands for now. We'll be ready to go in and start the rescue effort when the floodwaters subside, but until then we're pretty much tied down to a wait and watch position."

Kristen answered, "And the bad news is that we are showing still more rain heading into the area, based on satellite imagery coming into headquarters here. The storm doesn't seem to be moving anymore. It looks like that high-pressure system to the north is blocking any movement of Nisha for the moment, and if it does start to move, it looks like it most likely will move to the east toward the Malay Peninsula. We'd better start preparing Burma and Thailand for the possibility that the storm may be headed their way next, and even though the winds have let up for now, the rain is as intense as ever. It will likely cause major flooding wherever it tracks next."

"I know. We've been getting the feed from space and noticed that the storm isn't leaving the area any too fast. Without that high-pressure ridge, this storm would just about be finished by now. Any computer projections as to when it might break down or leave?"

"Maybe in another twelve to twenty-four hours we might begin to see some weakening, but only time will tell for sure. I know it's frustrating to watch so much suffering and damage go on and not be able to do anything about it, so you've got to do your best to keep morale high both in the Squad and in the local population. There's going to be a lot of work to do when it finally does break clear."

"We'll be ready Director; you can count on us."

"I know. You're all part of the finest group of people ever to serve their fellow human beings, and I know the world appreciates what you're doing out there. Anyway, I've got to check with Alex in Florida and see how that storm is progressing. Good luck, and I'll be back in touch soon. Bye."

"Thanks, we're gonna need it. Later."

"Is Alex still on the air?" Kristen asked Natalia.

"Yes Director, here he is now on audio."

"Alex, how's it going? We show the storm's eye approaching the coast of Florida and coming onshore in about two, maybe three hours maximum. Do you concur?"

"We're doing alright here in Disney World, although the wind is beginning to shake the building somewhat even where we are. I don't think we're in any danger though as Disney people have assured us that this building was built to withstand a direct hit from a tornado.

"As far as confirming what you're showing, we have indications that the storm is picking up speed and may be onshore in about an hour and a half, two at the most. We've done all we can at this point, and it doesn't really matter anymore when it makes landfall. In fact, we'd all just as soon get this over with and have the storm pass out of Florida. The waiting is the hardest part."

"I can understand. We're watching the scan cam from Miami right now, and the camera is shaking so bad we can hardly make out what's going on there. What's the highest wind you're clocking right now?"

"We're showing winds over four hundred kilometers an hour and an unconfirmed gust to five hundred. The rain is falling at close to seven centimeters an hour in some areas, and needless to say there is flooding everywhere from Miami to the Cape already. At the moment the eye is on a path that will take it south of Tampa, but with that much rain, they'll be flooded as well pretty soon. The leading edge of the storm is already into the Gulf of Mexico, and the edges reach from southern Cuba to Georgia. This is one giant storm!"

"Yeah, and to be that big and still so powerful is virtually unheard of in the weather world. It seems—""

"Director, sorry to interrupt, but we're getting a call from some Squad members near Lake Okeechobee that the water is being whipped by such high winds that it's beginning to spill over the banks, and the dikes are in danger of collapsing altogether."

"Can we get some equipment in there to shore up the dikes?"

"We can as soon as the storm passes over the state. The roads are underwater, and that will be slow to recede as there virtually is no place for the water to go at the moment. The equipment is large enough so that it can travel through underwater roads, but when it gets there the mud will be the greatest obstacle. Visibility is quite low, less than ten meters over most of that region. I have sent a message already for the nearest Squad members with access to the equipment we need to get on their way to the lake as soon as it is safe to travel. Wait—I'm getting a response already, and it seems that we have a crew in route as we speak. We happened to have a squadron stationed close to the lake, and they've got some earthmovers nearby and are on their way already."

"That's great! At least some things are going right. Changing topics for a minute, how many tornadoes have you recorded already?"

"We've seen thirty-five so far confirmed and maybe another nine or ten unconfirmed. Damage has been minimal so far, compared to the total destruction from the storm starting to come into our headquarters here. We've—"

"What happened?" Kristen asked. "He's off the air"

"We've lost contact with Orlando," Natalia said. "There is no signal coming in from the Florida headquarters at all. I'm trying to raise them on another wavelength, but we have no response at the moment. There just might be too much interference from the weather for the signal to come in." She was trying to lift Kristen's spirits against the pressure that she was obviously under.

"They'll probably be back on the air in just a few minutes." Kristen was hoping that they would be back soon, but had no direct knowledge about what was going on in Florida. Alex came back on the air for just a moment then disappeared again before a meaningful communication could be made. "Keep trying, and call me the minute that we make contact with Florida again."

"You've got it."

"Director, we're picking up a transmission from Helene," Elawa said. "She's on the main viewer now, I'm switching her to audio."

"Director, we've just received a report that a major dam just broke on another tributary of the Ganges, and just like last time, a wall of water is moving down the river toward the coast. We're expecting significant loss of life with this breakage and enormous property loss. It's more than we prepared for, because again the computer predictions didn't indicate that this could happen. It seems that the water is moving at about thirty kilometers an hour and is sweeping away everything in its path. The word is out, but most of the people downstream simply have no place to go. All we can do is assist with whatever help we can muster when the waters subside. The one good note we can pass along is that the rain is showing the first signs of abating. What do you show from space?"

"We show a break in the weather soon, and then another wave of rain. Then it looks like the rain may finally start to end. Computer projections show the storm weakening. It seems that some cool dry air from that high-pressure system in China is making its way into Nisha and finally the storm is starting to dissipate. Yet the worst may still be to come. With the lakes full, the ground saturated, there is no place for the water to go except where we don't want it to go. The projection for now is that the rain will be pretty much over in less than twenty-four hours, and with any luck the water will start to go down within a few days. Meanwhile we've got to do all we can to minimize loss of life and property damage. Is there anything else we can do from here to assist you?"

"I think we've done about all we can at this time. We've got standby units everywhere just waiting for the rain to let up enough for us to move out into the field. What's the situation in Florida?"

"It's bad, the winds are at record levels, and the rain is falling at very high levels throughout southern and central Florida. We're bracing for extreme damage both in Florida and in the Gulf coastal areas wherever the storm makes landfall next."

"I still can't believe that both of these storms are occurring at the same time."

"I think your earlier idea that they may be causing each other or at least linked to each other is a possibility and well worth investigating after this is all over. We've produced computer simulations indicating that the storms are a shadow of each other and that the reason both are so powerful is that they are centralizing a great deal of the atmosphere's energy into two points. They seem to be acting like a polarization of the earth's weather systems into two major disturbances. We'll know more after this is over and we've had a chance to study all the data from these two storms, but it seems that there is a correlation between the two.

"I'll call you back in a bit; we're getting a call from Alex right now and we've been out of touch with him for a while."

"Okay, talk to you soon."

"Alex, what's up? Good to have you back on the air."

"We've got another band of waterspouts moving on shore near Daytona Beach. We're switching to the scan cam we located there to see if we are able to spot them. Triple Doppler radar shows at least twelve and maybe as many as eighteen moving northwest at about fifty kilometers per hour, and they're only twenty kilometers off the coast at this time. They'll be hitting the coast in less than a half-hour at this pace."

"How far did you they are from the shore?"

"We're showing them about eighteen kilometers out and still moving about the same speed as they were when we first picked them up. By the way, you are recording everything from these scan cams, right?"

"Yes we are. Actually, we're recording all input coming into headquarters here from all the scan cams, remote weather stations, satellites, everything. When all the dust settles, we'll be able to recreate this entire storm—for that matter, both storms, from all the data that we've accumulated over the last two weeks."

"That's great, I want to see all this again when I have time to really watch without the adrenaline flowing like it is now."

"Eighteen minutes until landfall," Kristen said. "They're going right into the boardwalk area, which is a real shame, as most of that is still the original boardwalk that has withstood storms for the last hundred years or more, and this looks pretty ominous right now."

Everyone stared silently at the picture coming in from Florida. The room was inundated with the increasing sound from the twisters even though they were hidden from view by the rain and clouds. The roar grew deafening, and the director motioned for the sound to be lowered as the decibels increased. The minutes seemed to slow down in anticipation of the tornadoes hitting shore and seemingly destined to destroy the famous old boardwalk.

"Three minutes to landfall," Paul's voice said out of the silence and continued counting down to impact: "Two minutes … ninety seconds … sixty seconds. Forty-five—thirty … twenty … ten … five, four, three, two one, contact!"

Seconds later the boardwalk at Daytona Beach splintered into an explosion of debris flying high into the sky with pieces barely missing the camera recording the event. The path of destruction moved into the city, leveling a half-kilometer-wide swath right into the heart of the metropolitan area.

"Wow, that was terrible," Kristen said. "The only good news is that the band is so tight together. If these storms were spread any farther apart, the damage would even be more intense. Can you switch the scan cam further into the city?"

"We're switching now," Alex replied as the pictured changed to further into Daytona. The queue of tornadoes was still holding the same direction and forward speed as it ripped through homes, businesses, public buildings, parks, sparing nothing in its path. There was so much debris in the air that the storms were being obscured from the scan cams. As they watched, the funnels began to dissipate, and in another two minutes they were completely gone. It seemed that hours had gone by, when in fact only about eight minutes had elapsed. Eight minutes that caused hundreds of millions in damage, maybe some loss of lives at this point—but they could hope for only injuries, and the area had been mostly evacuated—the loss of historical sites, trees trashed into toothpicks, all in a matter of only eight minutes.

As they watched the debris crash back to the ground, Alex interrupted the silence with another announcement. "We've got more twisters on radar, this time in the south Orlando area. In fact, they're headed right at us!"

"Are you safe where you are?" Kristen asked Alex.

"I hope so. Disney officials said this place could withstand a direct hit from anything nature could dish out, but that was before this storm. We're moving into the maximum protection area and hopefully that will suffice. We don't really have any other place to go to at with the short amount of time we have until they're here. We're showing the storms only nine minutes away. They appeared only minutes ago, and we only picked them up on triple Doppler radar two minutes ago because the rain is so heavy that it's affecting all the radar in the region. It took us a minute in fact to confirm their existence. We're switching to the scan cam right outside of Disney here at Epcot. Can't see anything yet, but they should be coming into view pretty quick now from what we see here on the radar screen."

"We're picking them up now on infrared from space. They're less than ten minutes from you right now and they're moving—"

"Director, we're getting an emergency communiqué from Helene," Elawa interrupted.

"Patch it through now." Elawa already had Helene on the air, and Kristen turned to Alex and said she would be back in a moment.

"Director, this is Helene, do you copy?"

"We hear you, loud and clear. What's happening?"

"We're in big trouble here. The largest dam on the Ganges has just failed, and we're talking destruction on a scale not seen in modern times in this part of the world. There is frankly more water than can fit into the river ten times over. We're showing computer projections indicating up to a hundred kilometer swath being under water within just hours from the dam breaking. Well over fifty million people live in the area that could be beneath water very shortly.

"We're sending warnings out as we speak, but most of these people are without power already or have fled from lower areas into higher regions where they're out of communication, but I'm afraid that where they moved to won't be high enough when the big water hits. This is on top of the high water we already had in the river plus the earlier two dams that broke. We might see water as high as a hundred meters over flood stage. This could be just horrible for this region.

"We're picking up reports of the water moving in a wall down the river at about forty kilometers per hour, taking out everything in its path. Buildings are being swept away as if they were toys; so far not many people have been seen in the path, as this is an area that we did evacuate pretty completely before the storm hit. And this is an area with fairly steep banks. The region that will be most at risk is the plains area that is about twenty-three kilometers below this part of the river. The water will hit there in less than thirty minutes. We're going to need a lot more assistance than we ever dreamed of now. I'm afraid that we're going to be spread way too thin depending on the outcome of the next few hours."

"We'll get every possible Squad member into your region as fast as humanly possible, but I'd have to agree with you that we're spread quite thin now what with the typhoon, earthquakes, tsunami, and fires going on throughout the world. We'll need to put out a general call for any volunteers who can help anywhere in the world to assist in this disaster and the rebuilding that will be needed when this is finally over."

"That sounds like a plan. Has the Squad ever had to do this sort of thing before?"

"A few times when it was first formed, but mostly because there just weren't that many members in the beginning," Kristen stated.

"Director, the call has gone throughout the world, and the first volunteers are already showing up at Squad outposts around the globe," Markus reported.

"That's great. The response from the general public does always tend to be remarkable in times of true need. Also, Markus, continue the request for blood donors. With this kind of immense disaster we'll need all the blood that we can find, plus food, clothes, and of course clean drinking water. Start a general request message around the world for all the items I just mentioned and any transportation that can be recruited to move these items where they're going to be needed."

"Consider it done, Director."

"Thanks, Markus. Helene, we're already getting the first recruits showing up around the world in local unaffected areas. But it will take a little time to get people from where they are to where they're needed."

"That's okay, because nobody could get to where they're needed right now anyway. In fact, it may be several days before the water subsides enough for rescue help to get in there. We could use some help already in the other areas that weren't hit quite as hard. We need to establish triages for the wounded in several areas surrounding the worst disaster areas. Yeah, there's plenty to do for everyone whenever they get here. I'll call you back in a bit, Kristen; I need to check on some things here. Bye for now."

"Bye, Helene." Turning to Markus and Natalia, Kristen asked, "What's the status on Florida?"

"The tornadoes are reaching Epcot right about now; we're switching scan cams to see what's going on. Also we have Alex on the air again, after we lost them for a few moments."

"Alex, how's it going there?"

"Well, we're starting to shake a little here, but other than that you can't really tell anything is going on outside save for the pictures that we're getting on the cams. Quite a bit of destruction so far, as you can see: trees knocked over, buildings crushed, cars flung all over the roads. No reports yet as to the number of lives lost or the injured so far, but we did get warnings out to the entire area, and hopefully most people were already in some type shelter.

"Wow! Look at that power. The line of twisters is moving right through Disney World and the surface damage is intense. Watch that tornado heading straight at another scan cam … Here we are on line. The storms are hitting us now in their full fury—we're shaking like a leaf on a tree, but so far no problem. The funnels are almost through us here, and they look as if they haven't weakened as they move into the surrounding area. There are a lot of hotels, restaurants, miniature golf sites, and so forth in the area just ahead of the twisters. We're switching to another scan cam now as they move out of Disney World. It seems like they're starting to break apart as they move on; in fact, you can see that the storms on the left are dissipating as we watch.

"But wait, there is a call coming in from near Tampa … More tornadoes are touching down just southeast of Tampa–St. Petersburg. They're also moving in a northwest direction, a heading that will take them, if they continue on their present course, into the heart of the cities. There are a lot of elderly people in this region of Florida, and this could be disastrous if it hits with any punch at all. And at this moment it seems that it will. We've put out all the appropriate warnings, but we're dealing with a part of the populace who will be least able to respond to such warnings.

Alex paused. "Now more reports are coming in from near Cape Canaveral of tornadoes … and still another one from near Naples—three on the ground there … and still more up in Winter Park moving through the high school there. Reports from Winter Park are showing entire city blocks wiped clean of all buildings and trees. The tornadoes, as we expected, are becoming the main source of damage here in Florida. Communication is starting to break down in scattered areas around the state. We've lost communication with at least twelve of our reporting stations so far and others are faltering. It seems we're at the height of the storm right about now. What do you show from space?"

"We're showing the eye of the storm at almost the halfway point across Florida," Kristen said. "It is a pretty distinct eye now measuring about—let's see, eighty-five kilometers across. The winds near the center as measured with infrared Doppler radar is still an astounding 315 kilometers per hour on land. It didn't weaken very much at all crossing the peninsula, which means it will still be packing a full punch as it moves into the Gulf. We'd better get out maximum alerts to all possible landfall areas throughout the entire Gulf of Mexico and inland for likely floods that will follow the storm's path when it makes its final trek on shore."

After she put in the proper code for maximum full alert, the word was out all across coastal areas of the Gulf of Mexico. Most areas were already in maximum preparedness, expecting the worst after hearing the reports flowing in from Florida.

"The word is out," Markus said. "We posted a level-five warning. Anything else you want sent out, Director?"

"Not at the moment, I'm not sure what else we could tell them that they don't already know. Just keep a continuous flow of data on the storm's location to all interested parties."

"We've put the transmission of pertinent data on automatic now until the storm is over. We've included all the standard data of location, speed, direction, and so forth. There's about twenty-three different readouts being transmitted continuously."

"That sounds good, Markus." Kristen turned back to Alex. "We've put out all the appropriate warnings to everyone in an area two thousand square kilometers, taking in all of Mexico, half of the United States, and all of Central America. That's just about everybody in the northern half of the Western Hemisphere, except for Canada."

"Good, but we're still getting more reports here from all over the state; tornadoes are springing up everywhere. At last count we had fifty different areas reporting tornado activity, and probably some two hundred actual twisters have touched down with another three hundred funnel clouds sighted or picked up on triple Doppler."

Alex went on, "We're being pounded pretty badly right now. Scattered reports of flooding are coming in, also, along with further accounts of beach erosion along the coast, plus damage from the sustained winds of the hurricane itself. Unofficial accounts put the death toll at the moment above one hundred, most of them from a single collapsed apartment building in the tornadoes at Daytona Beach. It's not as bad as it could be though, considering the sheer number of twisters, the strength of the storm winds, and amount of rain falling.

"Here are more accounts of twisters hitting, this time, near Winter Haven where all the greenhouses are. They're breaking up and blowing away as we speak. There's flying glass everywhere; it's like a blender full of glass moving through the town. We're going to need quite a bit of help here when this through the region," Alex summed up.

"We have Squad members already in position," Kirsten responded, "just across the border in Georgia with helicopters, waiting for the all-clear to enter the area."

"Good to know. They can probably get into eastern Florida fairly soon as the storm moves west."

"We have Squad members already in the Bahamas too. The damage there is really bad. On some of the islands nearly every building is destroyed, and the loss of life has been pretty high. Hundreds are known dead at the moment, and thousands are still missing. It will days before we know the full extent of the catastrophe. There is still some hope that quite a few of the missing will be found; many had fled the region just before the full strength of the storm hit their towns. We'll know soon though, as rescue crews work through the region.

"I think we can have some Squad members move into eastern Florida now. With the threat of tornadoes diminishing in the eastern part of the state we're fairly safe to let the first Squad members into the region. We'll have them go to Daytona Beach first; that's where the worst damage has been reported so far in an area we can access. Any more news on the twisters?"

"The ones in Winter Park are gone," Alex said, "but they left quite a path of destruction. Winter Park High School was completely leveled, as was a good portion of the downtown business area. We took quite a hit here ourselves south of Orlando. Disney World has sustained enormous damage, especially in the Magic Kingdom. Most of the amusements and rides were destroyed, and the monorail was completely blown down throughout the park. With trees down, power outages almost universal now throughout Florida, communications out—the problems are mounting rapidly. We just hope this thing gets out of here while there's anything still standing in Florida."

"We agree with that, but then where does it go?" Kristen asked, speaking to no one in particular.

"Director, look at this lightning strike data coming in from Florida," Susan said as she was monitoring the statistics from the hurricane. The screen showed current lightning strikes as white dots on the screen. The state of Florida was almost white! Kristen just looked on in amazement; she had seen a lot of lightning screens before, but had never seen an area so completely white. Was it another record for this monster storm? When would it end?

CHAPTER THIRTEEN

The storm in Asia was winding down; finally. The winds had dropped to below one hundred kilometers per hour, and the storm was no longer officially a typhoon. The rain was still falling, but was way less than had been descending for the last few days. The wall of water from the broken dam had swept down the Ganges to the Bay of Bengal, leaving a gigantic swath of destruction on its way to the open waters and untold thousands dead or missing. This had added to the main problem, which now was the flooding: with dams broken, rivers over their banks, the ground saturated everywhere, there simply was no place for the water to go. Large areas of India and Bangladesh were underwater already, and more land would soon be inundated as the water upstream in the Ganges drainage area made its slow tedious journey to the Bay of Bengal. It would be weeks before the water would subside enough to even begin to think of things returning to pre-storm conditions.

Once the water was gone, there was the enormous task of finding all the victims, treating the wounded; getting food, water, medicine, and clothing to all those that needed help. Then the rebuilding task would commence. Literally millions of structures had been destroyed or damaged, millions of acres of farmland buried under debris washed down from upstream, and forests knocked over by the pounding wind, leaving many hillsides ripe for mudslides. For years erosion would be a constant threat. Then the problems of silt in the rivers would increase the chance for more floods from the monsoons when they started and would limit river navigation until the rivers could be dredged again and open up the waterways.

Infrastructure was completely gone in many areas, with every bridge, dam, roadway, and rail line damaged beyond use. This would take a colossal effort to rebuild. Already though, the major nations were pledging assistance to the effort. Many were offering money, materials, and manpower to the job at hand. Still others were offering medical relief teams, trees for replanting the forests, equipment for rebuilding the infrastructure, and personnel to run the machinery. It was a truly awesome outpouring of goodwill from around the globe to the people caught in this disaster. Without it, the region was going to have little chance of getting the calamity behind them. But then this was the way of the world in the middle twenty-first century.

Alex came back on with an update from Florida.

"Alex, how are you all doing?"

"We're fine, but the rest of Florida is hurting bad. We're getting reports in now from all across the state of major damage. Entire towns being wiped out, I mean every single building leveled, nothing standing! Reports of casualties are coming in too now, and I'm afraid it doesn't look good. Easily several thousand could be dead and tens of thousand injured. The sheer strength of the storm is the main cause plus the tornadoes caused a lot of damage. Flooding is also a problem but not the most severe one. Where do you show the storm now?"

"We are showing from the satellite images now that the eye of the storm is nearing the city of Tampa and moving in west-northwest, which would take it toward the Mississippi delta and New Orleans. We've put up extreme danger warnings in all of the coastal areas in the Gulf with a special warning for the area between Mobile, Alabama, and Houston, Texas."

"So when do you see it passing out of Florida?"

"Probably a little over one hour," Kristen answered, "but the rain and wind will linger for another twenty-four hours at least because the storm is so big. With any luck it will be out in the Gulf and away from Florida by tomorrow, but that means that it will most likely grow even stronger: the water in the Gulf is quite warm, even for this time of year. It's possible that Colin will grow even stronger than it is now, which would continue to put it in new record dangerous territory for winds, pressure, and rain.

"New Orleans is at special risk. It's mostly below sea level, and with the likelihood of twenty- to thirty-meter tidal surges, the entire region could be in serious trouble. Those levies were rebuilt to withstand category-three storms after Katrina, and they broke in the 2015 storm and flooded New Orleans, and they rebuilt them to withstand category-five hurricanes, but this storm might be too much for them.

"Anyway we've started evacuating the entire lower Mississippi River region and hope to have it empty by noon local time tomorrow. And as in Florida, the risk of tornadoes will be high. Small craft warnings are out for the entire Gulf, and air traffic is being diverted away from the region, along with all shipping lines. All the oil rigs in the Gulf that are still in operation are being vacated also. We've moved as much emergency equipment into place as we can, including front-end loaders, backhoes, dump trucks to haul debris out of the way, and other items, so that when this is over we're ready to get to work right away."

"Well, it sounds like we've done about all we can until this thing is over. How's the Asian situation?" Alex asked.

"The worst of the storm is over, but the flooding is just beginning because of all the rain and broken dams everywhere. It will be weeks before the impact of the storm is over and years to heal the damage."

"Director!" Natalia exclaimed peering at a monitor. "Quick, look at this."

"What the—is this what it appears to be?" Kristen asked in utter amazement.

"I don't know for sure, but it seems that the storm is splitting into two distinct storms with two different eyes. It may be that the size of the storm allowed for this situation to develop. The storm was almost a thousand kilometers across, and that may have been enough for it to split into two different and separate storms. I had noticed a wind shear coming into the middle of the storm a while ago, and it appears to have been able to split the storm in two parts. There must have been enough rotational energy in the outer bands to get a new eye to develop.

"And now it appears that the two eyes are starting to move in separate paths! One looks like it will hit land east of New Orleans, and the other is headed south of Houston … This is getting weirder all the time; so now one of the eyes is moving almost due west and the other is moving west-northwest."

"Can we verify this from space?"

"Stand by and we will have the latest pictures from above coming in over the monitor right about—now." As Natalia spoke, pictures from space began to appear on the wall screen, showing the two separate eyes of the storm moving away from each other on their individual paths.

"I wonder if this means that we have to give this new development its own name?" Kristen asked out loud.

"I think so," responded Markus, who had been monitoring the Asian situation, but listening in on this newest development.

"You're probably right. At the least, these new dynamics will set them on their own courses and make landfall in different places than we've been expecting."

"I'd have to agree with you on this one; it is now a new storm with its own set of upcoming problems!"

"So what is the next name on the hurricane list for this year?" Natalia asked.

"Let's see, we have the National Hurricane Center list right here," Susan said. "The next name is Daria."

"Daria? Well now, we have Colin and Daria. Does the National Hurricane Center concur with our observations?"

"Yes, Director," Susan went on. "In fact, we're getting their feed on the storm's new situation as we speak. They confirm two separate eyes and have labeled the second part 'Daria,' although they don't really know which part is first and which part is second. So they just randomly chose the section heading to Houston as the second part, and that is now Daria. They show Daria already well off the coast of Florida and Colin moving off the coast south of the Tampa–St. Petersburg area now. Daria is redeveloping over the warmer Gulf waters and is now stronger than the Colin in terms of wind speeds and low pressure. That's all the new information they have at this time. Their coordinates confirm our tracking of both storms, and they have issued the highest-level warnings throughout the entire Gulf region and flooding advisories well into the continent. They predict as much as a meter of rain will fall before this storm finally dissipates."

"Susan, make sure the hurricane center is receiving any additional data we might have from any of our different monitors."

"I'm on it. Plus our satellite images are all sent automatically to the center and other government agencies that have a use for such information."

"Thanks. Where is Colin now?"

"It's seventy-five kilometers south of the Tampa airport and moving slightly north of west-northwest. All planes, of course, left yesterday, and there is not much left there except for the terminal itself, the hangars, a few warehouses, and the fire station. Most of the area is evacuated according to our reports, but there are a few personnel in the fire station," Natalia replied.

"Can you raise Alex?"

"Sure, it will just take a moment. Here he is now."

"Alex, have you been following the newest development?"

"Yes, this is getting really bizarre. Two storms now to contend with; what will be next? Anyway we're showing an improvement in the situation here. Winds are letting up a little, and the rain has slowed for the most part, although we're still getting the occasional squall line moving through with heavy rain. The reports from Tampa are bleak. They're getting pounded with both very strong wind and heavy rain. The damage will very likely be quite severe.

"There is good news though, and that is, the storm has picked up forward velocity and will clear the coastline sooner than we first thought. That may help with some of the rain problems and the flooding that will follow, but probably not with the wind damage and tornadoes that are occurring right now. It should be off the coast within the hour at the present rate. Let's see, that would have the winds tapering down in about … maybe three to five hours. We could have the first relief crews in position near there and move them in as soon as the wind drops to a relatively safe speed. We have people in the eastern portion of Florida already starting with the rescue effort.

"The first reports coming in from the Miami area aren't too bad; with only a few dead, maybe one or two hundred injured, and property damage moderate. As you move north the damage gets very bad very quickly, and the number of dead and injured goes up as well quite rapidly. The worst area at the moment is around Daytona Beach because of the tornadoes that passed through there. Others include Orlando, especially the southern half of the metropolitan area, around Lakeland between Tampa and Orlando, and of course the Tampa–St. Petersburg metro area.

"North of Orlando," Alex went on, "the damage begins to subside gradually until about the Georgia border, where there is very little destruction other than the occasional sign blown down or a few shingles knocked off a roof. That area missed the storm's major winds, although they did receive some substantial rainfall, but not enough to cause any more than very insignificant flooding. We'll be into full-scale rescue across the board here shortly, so I'd better check readiness status throughout the region. Anything else you can tell us now about the situation?"

"No, you're as current as we are for the moment, but if anything changes, we'll give you a quick buzz; otherwise we'll check back in with you in say an hour."

"Roger that, bye for now."

"Director, Helene is calling, and I'm switching to the big screen," Susan stated as the image of Helene appeared.

"Director, the situation is getting completely out of hand along the Ganges. Several dams have broken upriver; unconfirmed reports put the number as high as ten or twelve, and all that extra water is reaching the river at basically the same time. Reports of the channel overflowing to twenty kilometers across are now coming in, and it appears to be getting wider.

"Needless to say, there also reports of huge loss of life. Some areas were more prepared than others, but the vast majority of the people didn't even know it was coming; The rain has let up, giving the impression that the worst is over. Some were even heading back into their homes closer to the river when the water hit. We've tried to get the word out, but communication is still limited at best and mostly nonexistent. All power is out throughout the region, so only battery or solar powered radios are operational, and it's night here, so solar is out for the moment. Any ideas on what else we could do? Although at this point, very little time is left before the water hits some densely populated areas."

"Have you tried planes flying over the riverbanks?"

"We did in the daytime, but with night, most of the people have taken shelter where they can find it and aren't watching the sky. We were doing both a banner flying behind the plane and loudspeaker warnings, plus it's too dangerous flying in the dark as low as you have to in order for the loudspeakers to be effective."

"I see your point on that," Kristen said. "Let us work on it for a bit, and we'll see what else we can come up with to warn people. It must be nearing dawn though, is that right?"

"Yes, in about two hours, and we'll get the planes back up as soon as we see treetops again, but the pilots report that it seems to scare the people more than warn them. They say the people don't even look when they fly over and just keep heading toward the river, or nearer the river they make no attempt to move to safety."

"How's the rest of the area faring?"

"Well for the most part everywhere else is doing pretty good considering what they've just gone through. The rain is nearly over in most southern areas, and the wind is down to almost nothing across the entire region. The only life-threatening situation at the moment is the flooding; it's widespread, but is the most serious along the Ganges River. That's where the loss of life will most likely be the worst. The latest satellite picture shows the storm breaking up, which is extremely fortunate, as any more rain would have caused a disaster of biblical proportions for hundreds of millions of people. The floodwaters are expected to peak within twenty to thirty-six hours and could reach as much as thirty meters over flood stage. The good news is that after the crest passes, the water level should drop off dramatically, since when the dams broke, all the water came at once. We could be pretty much back to normal within about two days as it looks now. By the way, how is the Atlantic storm progressing?"

"Eastern and Central Florida were hit pretty hard, and Tampa still is getting battered"—Kristen paused for breath—"well ... it's bad there. The wind was as strong as any wind ever recorded in the world, topping 388 kph. Buildings were heavily damaged or destroyed, and the loss of life is still unknown although the preliminary reports are already grim. It will be quite a few days before we know the real number of casualties.

"And now the strangest thing has happened; the storm has broken into two separate storms, each with its own eye and now its own path. One is heading south of Houston, and the other is moving toward somewhere between Mobile and New Orleans. Daria is the name of the new storm, and Colin remains the name of the original storm. There was some confusion as to which eye was the first and which was the second, but it doesn't really matter; all that really matters is that we have two storms to deal with now."

"This is too weird ... When do you project next landfall?"

"Within forty-eight hours it looks like both of them will have hit land. Colin, which is headed toward Alabama to Louisiana, is on a course and speed that will put it ashore around thirty hours from now, and Daria, which is headed to the Houston area, will be onshore between forty and forty-eight hours from now."

"Give Alex my best; we'll all need to get together after this over and trade stories. Well, I'm getting another call from along the Ganges, so I'll call you back in a little bit. Out for now."

"Keep us informed if anything changes; talk to you later."

"Director, Alex is on the other screen for you," Natalia said. She stifled a yawn, showing just a hint of the exhaustion that all of them were beginning to feel. It had been quite a while since anyone in the control room had slept, and it was looking like it would be at least another two days before they would be able to get any substantial sleep.

Most of the work they were doing was to act as a clearinghouse for all the information pouring in from everywhere and get it to the parties who needed the information. Input from hundreds of sources was feeding into the computers housed in the Squad's headquarters, which in turned was analyzed, dissected, and used to project every possible outcome anyone might need, and those findings were then fed to those that needed the data. It was an awesome task and had to be watched continuously as things changed.

Alex's face appeared on the screen looking as tired as they all were at headquarters. Even Silverton seemed to be tired, but at least she was able to sleep, which she decided to do now on the warm computer. Her little purr could barely be heard over the whirling, beeping, and clicking of the computers.

"Alex, what's your status?"

"The worst is over for Florida; the eyes have both passed over the coast and into the Gulf. The second act is over for these storms; now for act three. Squad members are beginning to move into place by helicopter and transport planes all the across the state. The first reports are bad for the middle of the state. We have over fifteen hundred confirmed dead already, and thousands are injured, but we have triages in place and have begun to assist the injured.

"The good news is that twenty-three were rescued from a collapsed building in the Daytona Beach area, where they were treated for minor injuries and released. We've had some other good fortune too across the state. Two children who were separated from their parents were found safe in a car that they crawled into during the storm. An elderly couple was taking a drive and hadn't paid attention to all the warnings when they drove into a flooded road and were swept downstream, but they were washed up on the bank and managed to escape to safety. More stories like these are coming in to headquarters here, along with the downside of the storm's toll. But when you consider how many people live in Florida—well, it could have been a lot worse.

"We should have more complete reports within twenty-four hours as Squad members get to every area and finish the rescue effort. We're getting good support from local police, National Guard members, and just everyday people volunteering to help out wherever they're needed. One thing we do need is more blood donations. We've had a lot more injured than we had anticipated, and we're short of blood supplies."

"We'll get the word out for more donations and get the blood to you ASAP," Kristen said, and with a nod to Markus, the request for blood donations went out across the Western Hemisphere.

"Thanks, I'll relay it to the cities where we need it most. And the sooner the better, we're running low. Also we'll need more drinking water pretty soon. Some of the supplies that we had in place for after the storm were destroyed by tornadoes and the wind gusts that ripped the state."

"We'll take care of it right away. Anything else?"

"Not at the moment, but we'll keep in touch. I'll report back in another hour and update you on the Florida situation."

"Talk to you in a bit," Kristen said and then, turning to Markus and Natalia, asked for the current status of Colin and Daria.

"Well, not much has changed in terms of velocity and direction since they left Florida," Natalia began. "We're still showing the approximate same time and same landfall for both storms as of right now, although Colin keeps speeding up and then slowing down since it left Florida. But so far it has averaged what we first projected for it.

"We've sent all the current info to the appropriate officials and put out the maximum level-five warnings, predicting extreme damage likely for the entire Gulf region, and flood watches well into Mexico and the United States. We're showing the potential for major flooding in the Mississippi basin as far north as St. Louis and all areas south on virtually every stream, creek, and river in the area. We're advising evacuation of any site near water, everywhere between the Rocky Mountains as far north as Colorado and the Atlantic Ocean. This is a very large area that is at risk from these storms. Also, Director, the computer is showing a 20 to 30 percent probability that these two storms will merge into each other again, perhaps over Texas. Then will it be Colin or Daria if it does do that?"

"I don't know; it's never happened before. I guess we'll leave that question to the National Hurricane people if and when the time comes."

"That's what we thought also. Is there anything else you want us to do, Director?"

"Not at the moment other than keep tabs on the blood supplies as they become available and get them to where Alex says they're needed as soon as possible. Other than that, it sounds like we're on top everything for the moment until the storms make landfall again. I'm going to my office for half an hour or so to do some bookwork until something changes, but you know the drill ... call me if anything changes."

"You've got it, Director," Markus replied, and Kristen left the room. Silverton woke up when she started out of the room, quickly jumped down from the warm computer, and followed the director to her office. Her tail was straight up in the air like something important was going on in her world.

"Susan, now that we have a lull for moment, have you gotten any new scores on the World Cup games played in the last few days?" Markus asked.

"No, but I'll pull them up on the Internet sports section. It will just take a minute. They'll be up shortly."

"I'm afraid that will have to wait," Natalia interrupted. "Get the director back here; we've got a new problem."

Kristen never even reached her office before a call from Susan over the loudspeaker system summoned her back to the control room; she went straight to the computer terminal everyone was standing around and looked at the pictures. Silverton continued to her cat door and went outside for a moment, but she too was back in the control room just minutes later.

"Switching to the big screen now," Markus stated, and the image of tornadoes ripping through the city of Tampa dramatically appeared. He went on, "Director, we're showing some thirty to fifty twisters on triple Doppler radar. There was no warning; they just dropped out of a massive squall line that was moving toward the coast and would have been the last of the major squalls to leave Florida. We've sent warnings out, but I'm afraid there wasn't time to warn a lot of people. The other bad news is that people had started coming out of their shelters as the wind and rain had subsided for the most part."

"Which direction are the funnels headed?"

"Northwest. The squall line is moving counterclockwise around the eye coming up from the southeast. They should be over water within fifteen minutes, but they could do some major damage before they're out of there."

"Patch in Alex."

"I'm with you already, Director. We've been following Markus's report from the start. We've got Squad members on the way to assist the ones already on the scene. Anything else we should know?"

"You're as current as we are; they just appeared a few minutes ago and—"

"Sorry to interrupt," Markus said, "but more are appearing. It seems this squall line is extremely conducive to tornado formation. Some are materializing for just a moment and then disappearing, while others form and persist. I've never seen anything quite like this before. Has anyone else here seen this before?" he asked to no one in particular.

"Not to this degree," Kristen responded. "It's pretty common for these lines to produce tornadoes, but not this number and not so fast. Plus do we have any data as to when the first ones emerged? Were there any before Tampa?"

"None that we show. It seems as if the first ones were these that we're watching right now. The conditions must have been favorable right here in Tampa and not before, which is a little strange, because what changed from say twenty minutes ago to right now?"

"Well we've recorded all the information and when we analyze this later, maybe we can tell," Susan stated, "but it is rather peculiar that they only formed in Tampa. Let's hope people are listening to their radios and heed the warnings. How many do you show now?"

Markus answered, "At least forty. They're so close together that it's hard to differentiate individual twisters; however, the computer is fairly accurate in its count, plus or minus 5 percent. Add that to the fact that some of the tornadoes have a lifespan of less than a minute, and you can understand why it is hard to pinpoint an exact number at any one time."

"Markus, are these storms still on target to be off the coast and over water in a few minutes?" Kirsten asked.

"Yes, we're lucky in that regard," Markus responded. "However, check on these pictures we're picking up from downtown, buildings being ripped apart, glass flying around like ice in a blender, and the few trees still standing are going down violently. This looks really bad ... real bad. The first twisters will be off the coast in eight minutes, and the last in twelve minutes at current speeds. The worst should be over by then."

"That's good news. Do we show any other squall lines over the state with this capability?"

"Not at the moment," Susan stated watching the Doppler radar screen over Florida. "But then we didn't show this one until it started to sling down twisters, so we can't say for sure."

"Alex, I know you know what to do; sing out if we can assist in any manner that the Squad needs. We'll be back in touch soon. Out for now," Kristen said and, turning back to the big screen, acknowledged a call from Helene coming in from India. "Helene, how it's going?"

"Well the crest of the flood should be occurring within three hours, providing no other dams break and the rain continues to diminish like it has for the last few hours. The flood stage will be exceeded by somewhere between twenty and thirty-five meters. That's a lot of water when you consider how flat this land is at this point on the river and how far this flat land extends away from the water. We're talking thousands of square kilometers that will be underwater before this is over. Any crops that were left standing from the storm itself will surely be wiped clean off the face of the earth now. It will be at least a year before this area will produce crops again or sustain much life at all for that matter.

"We've got Squad members there on the scene now, following the crest downriver and commencing rescue efforts as soon as the water begins to subside. The effort is being hampered by the fact that there is literally no way to get into some of the regions. All roads are gone, there is nowhere to land a plane of mud and debris everywhere, and the water is moving too swiftly for any boats. Helicopters are being used where they can, but we don't have many at our disposal. So we just pick up people and get them to the closest dry land and return as quickly as possible. It's working, but it's slow. We're pushing makeshift roads into the stricken areas as fast as we can get road base in place. This will get us closer to the flood victims by tomorrow morning, then we can get supplies in here, medical personnel and so forth."

"That's about all you can do for now until the water passes. How many people do you have ready to go?"

"We've been getting new recruits coming in continuously for several days now, so I don't have an accurate count at the moment. However, we are hearing from our regional supervisors somewhere in the neighborhood of twenty thousand Squad members in the subcontinent area with several thousand more in outlying areas helping with the flow of supplies into the region. That's about all that we have the support systems for at this point. There just isn't enough infrastructure left to handle the flow of food, water, and the rest to keep the Squad members going. The last count we received here about an hour ago showed seven hundred bridges damaged, almost every road in the area of the storm impassable, and practically every single airport out of commission, although some can take in a few smaller planes and helicopters."

"We'll see if we can't round up some more earthmovers and help get those roads built as soon as possible," the director told Helene and then glanced at Paul who was in charge of acquisitions for the Asian storm area. He nodded back, acknowledging the request, and went right to work finding the needed equipment. Paul had been busy, working to find extra available equipment wherever he could and line up the transportation necessary to get the gear where it was needed. He was running out of options as almost all available infrastructure repair machinery was already being used and had a waiting line ready for the use. But he did manage to find a few extra earthmovers in southern Australia and locate a transport plane that could deliver them within twenty-four hours. He told Kristen, and she relayed the information to Helene.

"Kristen, thank Paul for his effort from all of us here. Well, I need to check on some things here and see how we're doing. I'll get back to you soon before the river crests and update you on the situation. Out for now."

"Sounds good, let's say an hour from now, okay?"

"Okay, bye."

"Thanks, Paul, from Helene."

"No thanks needed, it's my job."

"I know, but a little gratitude every once in while never hurts. Now where did you find these earthmovers?"

"Australia, near Melbourne. We found two military transports that were large enough to get them airborne right away."

"Keep looking for some more; you know they'll need all we can find."

"You got it, Director," Paul answered.

"Natalia, what's the latest on the line of twisters in Tampa?"

"They're just about off the coast now; two or maybe three are left onshore, and they're leaving. Here's the radar screen now ... the last one is over water now. Squad members are moving into the city from all directions now; the rescue effort is in full swing. The first reports are coming in ... We've got minor casualties so far, a lot of injuries especially from flying glass. Our people are setting up a triage to treat the injured. We are getting more reports from the downtown region ... a building that was completely destroyed ... hundreds live in the building, but there's no report as to how many were in the building when it collapsed. Okay, now an eyewitness at the scene says there were quite a few ... The first of our people are approaching the building, and they can see some bodies in the rubble ... They're going to need equipment to remove the debris and get to the bodies—and hopefully some survivors."

"Let's hope," Kristen answered. "Natalia, where's the closest equipment that could pull off the collapsed building?"

"I'm checking ... Here it comes on screen—just outside the Tampa city limits, maybe six miles from where it needs to be. We've got people on it now to get it into position. Should be there in say fifteen to twenty minutes."

"That's good," Kristen answered back. "Any other tornadoes in Florida that we should know about?"

"Not at the moment. It looks pretty clear at the moment. The rain is subsiding, and the winds have dropped off to barely gale force across most of the state. The bad news is that the storms, both Colin and Daria, are intensifying as they move into the warm waters of the Gulf. They are both near the strength the first storm had when it came ashore. This is a little confusing … first one storm, then two storms. Anyway, both of the new storms are getting stronger. We show them coming onshore some time tomorrow and the next day. As soon as we have a more definite time of landfall, Director, we'll let you know. Anything else?"

"Not at this instant, thanks."

"Director, look at this," Elawa said, and the director came over to check out the screen she was monitoring. "Good thing this is over water and not land. That line of twisters has become almost a solid band of waterspouts ten kilometers wide moving through the ocean. It's a good thing that no ships are in the water or they would surely be torn up."

"Wow, I've never seen anything like this before; it sure is lucky that it left the Tampa area before becoming this intense. We would have had damage beyond comprehension, from what I'm seeing on this screen."

"Yes, Director, this is very unusual; look at the Doppler on this line. It shows there is no area at all between the twisters—it's as if they have become the same tornado. One giant waterspout running some thirty kilometers per hour across the ocean. Save for the grace of Nature herself, this would have been over a populated region of Florida."

"That's for sure. Nothing in the way of this line of storms?"

"No, only open water. We're certainly lucky on that account."

"Can you raise Alex please, Elawa?"

"Certainly, Director. Here he is now on the main viewer."

"Alex, are you following the line of tornadoes that just left Florida?"

"No, we've been tied up trying to get people rescued, what's going on?"

"They've basically become one waterspout ten kilometers wide moving at thirty kilometers an hour over the water. It is an amazing sight on the radar screen. Anyway, we don't want to keep you from the work at hand; we'll show you all this later. I think we're really lucky, though, that this didn't occur while they were over Tampa."

"You can say that again. It's bad enough from what did happen, let alone something like you're describing. I need to go, I'll get back with you when we know more about what transpired here when that line of twisters hit the city."

"Catch you later," Kristen replied and then asked Paul how it was going finding equipment for India.

"We found one more earthmover, and it's on its way as we speak, but that's all for now. All the others are being used in other rescue work, either in Thailand with the tidal wave or China with the earthquake and so forth. But we're not giving up and will keep looking."

"That's the Squad's spirit that makes our work worthwhile," the director stated, obviously proud of her colleagues, who were certainly exhausted, but no one ever complained or even mentioned it. It went with the territory. The Squad was a well-trained group of individuals, trained to withstand all degrees of adversity, and the people in this room were the best of the best, the crème de la crème. Situations like this were exactly what the years of training were all about: being ready for these extreme circumstances. Kristen knew that she could count on everyone in the control room to be there as long as they were needed.

"Susan, what are the latest landfall points for each of the two storms?"

"Nothing has changed for now; the targets are still the same," she answered. "However, Daria's forward speed has increased, and we've estimated that it will hit near Houston from four to six hours earlier than originally forecast. We've sent the word out to all affected parties. Colin hasn't changed speed at all and is still on course for the Mississippi panhandle to New Orleans. We're updating every fifteen minutes from now on until this is over. Anything else you want us to do?"

"Not at the moment. If anything changes, let me know, of course," Kristen replied and Susan nodded.

Indira, who was monitoring the rest of the world's emergencies, spoke up when there was a break in the dialogue, "Director, thought you would be interested in the rest of the world situation. The fires that have been burning are pretty much under control except for the one in the western United States. All the others are either out or at least contained and expected to be out within twenty-four hours. The floods in Colombia are somewhat better, although the number of lives lost is still unknown as many of the victims are buried under meters of mud and will probably never be found. The good news is that no new dams have broken for over forty-eight hours now, and the rain is abating quite a bit. It looks as if the worst is over, and now the main concern is safe drinking water and of course medical supplies."

"What is the water situation?" Kristen asked.

"We've got equipment on the way to set up purification centers and several planeloads of bottled water to tide over the people until we start producing drinkable water. Also we have purification tablets being airdropped in outlying areas along with food and medicine. People can boil their water also, but fuel is low everywhere, power is out pretty much over the entire region and it's so wet that wood is hard to get to burn. The situation should approve shortly as some power is scheduled to be back on within twenty-four hours, barring any further problems."

"Sounds good, keep us posted, Indira."

"Will do. By the way, the fire in the United States has leaped the fire line that they hoped would control it, and now it's in a whole new area of forest. They're building the next fire line and plan to start a backburn from that location as soon they're ready. This line is pretty close to a national park, and there is real concern if they can't stop the fire here, huge areas of virgin forest will be at risk. They will send word as soon as something new develops."

"Let's hope that works. There aren't many areas of pristine forest left in the world. It would be a real shame to lose such an area. How many planes are involved in the effort?"

"About fifteen. For a while they were flying back and forth between several fires; now that the others are out, they can concentrate on this fire. We've got five other planes that were damaged by high winds in a thunderstorm several weeks ago, and they are just about repaired. Hopefully they'll be in the air by tomorrow. At least that is the forecast at the moment. Again they're keeping us posted on events as they develop."

"Where were those thunderstorms again? I remember hearing about them, but I don't recall exactly where they were."

"Let's see—here it is, they were just outside Salt Lake City. It was a pretty freak thunderstorm, with possible tornadoes, but the winds were confirmed at over two hundred kilometers an hour. It caught the planes on the ground and not secured against winds that strong," Indira stated.

"Thanks, Indira."

"Director, it's Helene on the screen," Natalia inserted as the scene from India came over the airwaves.

"Helene, what's the flood's current situation?"

"Well we're quickly approaching the crest of the flood. The water is over thirty meters above flood stage already and is expected to reach about thirty-five meters at the peak, which is the high end of what we originally predicted. The flood is stretching some 175 kilometers on either side of the Ganges. The area will take years to be productive again, as the amount of debris that has washed over the land will take a long time to clear away. The good news is that once it dries out, the soil will have benefited from the silt that has washed down from the mountains. We've got a plane overhead now and are switching the cameras on for you. Here they come now; as you can see, the water is covering a massive area, all the way to the horizon in some directions."

Helene continued, "The pictures tell the story. When you hear about something like this, you never quite imagine exactly the scope of the disaster. Then you see the pictures and … nothing prepares you for what your vision brings to your senses."

"I know what you mean," Kristen replied.

"It looks like the ocean—a dirty ocean with trees floating along, the brownish color from all the soil being swept away, people's homes, their possessions drifting toward a watery grave in the sea, all memories gone with them; but new memories will be created, and life will go on."

"Yes, that's for sure," Kristen said. "How long did you say until the crest reaches this area?"

"Any moment now and for sure within the hour. We'll keep the scouting plane overhead while the river crests so that we can know when the worst is over and get Squad members into position quickly. We have units already upriver searching for survivors. Numerous small islands have been created where many people and animals are trapped. These are the areas that we are concentrating on first. After that we can search the peripheral regions for any more survivors.

"We'll have to wait for the water to subside before we can look for bodies buried in the mud. There will probably be a considerable number that we never find at all. We just don't know how many were in the flood's path when it hit; plus most likely entire families or even villages were wiped out so there won't be anyone to report someone missing."

"Helene, Alex is on the other screen. We'll keep you on the split screen, and I'll be back shortly."

"Understood, Director."

"Alex, what's up?"

"We've gotten into the mid part of Tampa and we have an entire block demolished. Our first examination of the rubble looks really bad for the inhabitants that used to live here. We found thirty-some bodies lying around the edge of the building that were hurled outside when the twisters hit. There are a few survivors who have stated that the buildings were all full when the tornadoes ripped through here. Our best guess at the moment is that there were close to five thousand people living in these apartments. It was a series of twenty-story apartments and condos; the residents were mostly retired, but there were some younger families with children. We've got a full crew on it and heavy equipment on the way to start the removal of the collapsed buildings. Also there are scattered other cases like this all over the city, but this is the worst so far that we have found.

"Our death toll in Florida now stands at thirty-five hundred confirmed, and there are reports of another ten thousand people missing. The number injured is at nearly twenty thousand so far, mostly from flying glass and debris. We are expecting the worst and have sent out an urgent request for more blood and medical supplies. The first ones are beginning to enter the state now, and we should have them where we need them soon."

"When do you expect to have the buildings cleared away to see if there are more survivors?" Kristen inquired.

"It will take days to complete the job; these were large buildings and they are pretty much leveled, plus there is noplace nearby to take the debris. The streets are full of litter, toppled trees, smashed cars … It's real bad downtown, and we haven't even gotten to the coastal regions yet where we understand the storms became rather intense."

"Yes, we can confirm that here. We were monitoring the storms as they went through the city, and they seemed to grow substantially as they departed the land, where as you know they have become one giant waterspout."

"We have Squad members in route to the coastal region of the city, but the going is slow as there is so much wreckage in the streets. They should be there soon and when they arrive we'll get you an account of the situation"

"Let us know if there is anything else you need, Alex. We're all stretched a little thin, but we could probably find something somewhere if you need it."

"It's just going to take time now to dig through all this mess, but we've got the best crew on the world to do it."

"Roger that; talk to you in a minute. The floodwaters on the Ganges are cresting any moment now, and we're monitoring that area also. It's bad there too; thousands may be lost in this out-of-control river."

"We can all understand their loss. When will this all be over with?" Alex sighed, showing exhaustion and the frustration of not being able to help.

"Helene, do you copy?" Kristen asked.

"Loud and clear. The river just crested and is slowly starting to recede. It will be at least a day, maybe two before the water goes down enough for us to really get in here and get to work. We've got helicopters flying over the flooded area pulling up people we find and airlifting them to dry ground. Other than that it will be a while before we can really get in there."

"That's probably correct. We don't want to put any of our people in needless risk, and God knows there is enough risk out there without us adding any more!"

"We're about to launch boats upriver behind the most turbulent waters and have them search for anyone or anything that needs help. The water is calming down somewhat twenty to thirty kilometers behind the crest, and that's where our watercraft will remain as the crest moves toward the Bay of Bengal. The crest should reach open water in about fifteen hours; barring any unforeseen developments, that should be the worst it's going to get from this storm. Then the rescue effort can really get under way. How's the effort in Florida going?"

"It's pretty bad," Kristen said. "Thousands may be dead, and the number of injured is staggering. The storm was extremely intense, and tornadoes ransacked large sections of the state, especially

Daytona Beach and Tampa. Then the line of twisters moved out over the Gulf of Mexico and became a solid waterspout ten kilometers across, and they're still going at this time. We just hope they dissipate before they reach land again."

"Yeah, that's for sure," Helene responded. "Well, give Alex our best, and I'll be back in touch soon."

"Okay."

"Director, you need to come and see this." Markus was clearly excited about what he was staring at on the screen.

"What's up?" Kristen inquired.

"Look …"

"Is that what I think it is?"

"I do believe so. That line of tornadoes has become its own hurricane!"

"How does that happen?" the director wondered out loud.

"I don't know for sure, but I do know that I've never seen anything quite like this ever occur. They seemed to break apart from the rest of the storm as they were trailing in a long squall line that extended at least a hundred kilometers behind the main portion of the hurricane. Then right as I was watching the satellite imagery, the line of waterspouts began their own larger circulation, and an eye formed after that. Well, what do we call this one?"

"Check the list and see what it says—"

"I found it: *Eric.* This is a record, by the way; no other storm as ever spawned two other hurricanes from itself. We've had one create one before, but never two."

"Just one more record for this storm," Kristen declared, and then asked, "What direction is it heading?"

"Well at the moment … let's see … north-northwest. That's more like the panhandle of Florida into Alabama. Look at this picture of the Gulf of Mexico from space," Markus said pointing at the computer screen. "I don't think there's a square kilometer that's not covered by clouds, and see here, here, and here are the three eyes of the storms."

"Have there ever been three hurricanes in the Gulf of Mexico at the same time before?" the director inquired.

"I'll check that out …" Markus responded. "It seems not. There once were three tropical disturbances, but never three hurricanes at the same time. Now we've got three different landfall probabilities to contend with. We've sent out the warnings to the new sites and have updated the other current targets as soon as we had the data from the newest storm. Daria is still headed toward Houston, Colin toward New Orleans, and the newest one, Eric, is aimed toward Mobile at this point. Projected landfall times run from fourteen to twenty-four hours now for all three storms. Daria has picked up forward velocity and may now be the first of the first two storms to reach land. Already Houston is reporting gale force winds in advance of the hurricane, and tides are running some two to three meters above normal already. Colin is heading pretty much toward New Orleans, where winds are also starting to increase, and tidal surges are reaching as high as three to four meters. Eric is tracking toward Mobile as we reported earlier and will reach there within eight to ten hours. Eric is the weakest of the three storms and has the shortest time over water to develop before making landfall at its current velocity and direction. But Eric is still a dangerous storm. It is likely to do severe damage as it comes onshore."

Markus went on: "Tidal surges are reaching Mobile already; in fact, we are now reporting tidal surges around the entire Gulf. The water throughout the Gulf is pretty turbulent, as you might imagine. We've had small craft warnings out for days. Most little boats have found shelter, and all large ships left the region with the first threat of the storm a week ago. All oil rigs have been completely evacuated, as have all coastal regions from Florida to Mexico, and low-lying areas throughout the southern states along every river and stream have also been abandoned. At least we've had plenty of time to prepare for the coming onslaught.

"Report from Houston coming in … winds have just gusted to hurricane strength at the airport. A few waterspouts have been picked up on triple Doppler off the coast. The rain has begun to fall in earnest along the coast and is starting to move inland. That's the latest from Houston and the Gulf update," Markus concluded.

"Who's the highest ranking Squad member in the Houston–New Orleans area?" Kristen asked.

"I've got that information right here," Natalia piped up. "Her name is Debra Holbrook. She's from California and has been with the Squad for seven years, mostly on assignment in Asia and the Africa. Speaks twelve languages and is a well-known photographer also. She's at the headquarters set up in between Houston and New Orleans. Stand by and I'll raise her on the screen." In moments she continued, "Here she is now."

"Greetings, Debra, I'm Kristen Verba, Squad Director."

"Yes, I think we met at a Squad seminar on rescue techniques of natural disasters about seven years ago."

"Yes, I recall that seminar. Sorry I didn't recognize you, but it has been a long time. But it is good to meet you again."

"Same here, what can I do for you?"

"You are the senior Squad member in the Houston–New Orleans area and will be in charge of the rescue and rebuilding effort after these storms make landfall. Are you current with the situation?"

"Yes I am. It's really unusual about the two other hurricanes being spun off of the first one, isn't it?"

"Yes it is; in fact, it's the first time this has ever happened to the best of our knowledge, and it's not over yet. Who knows what will happen before it's done?"

Debra asked, "What are your current projections for landfall?"

"Watch your screen; we're sending you all the current data we have on all three storms. As you can see, we expect the first of them to hit land within eight hours. That would be Eric at this moment, which is heading toward Mobile. The others will follow closely after that. Houston is already picking up gale-force to hurricane-strength winds, and New Orleans is getting three- to four-meter tidal surges by now. The next twelve to twenty-four hours will be the worst for the entire Gulf region."

"We're ready; the entire area is evacuated along the coast and all low-lying areas too. We've got plenty of supplies on hand, including food, water, medical, and so forth. We're as ready as we can be. I personally find the waiting the hardest thing about these situations."

"That's for sure, but it sounds that you're all ready; good luck to everyone," Kristen said. "We'll be back in touch every hour from now until landfall. Then we'll go on continuous communication until the storm is over."

"Thanks, I think we're all going to need it. We'll talk to you in a while. Out."

"Bye, Debra, talk to you soon."

"Director," Natalia said, "Helene is on the screen; she needs to talk to you right away."

"Helene, what can we do for you?"

"Director, we've got a new problem; another dam just broke, and we've got another wall of water coming down the pike. This is on a different river than the Ganges, which is fortunate in one way, but opens a new area to flooding. This is the twenty-third dam we have confirmed broken, and the rumors are that at least another eighteen have been damaged to the point of releasing water."

"Sorry to hear it. What can we do to assist the situation?"

"Not much at this point; we've sent Squad members into the area, and they will report back to us as soon as they know something. The river runs through a heavily populated region of India, and we've sent warnings out throughout the affected area."

"Well if anything changes that we can help, let us know."

"We'll keep you all informed, but the best thing we can do now is to get people out of the way as fast as possible. Then we can set up rescue units to get out anyone who's trapped. We're getting our first reports from the Squad members on the scene. We're switching them on live now so you can hear as we do," Helene stated as the sound of helicopter rotors came over the air.

"Headquarters, do you copy?"

"Loud and clear, what do you see?"

"We're over the headwaters of the flood wall now. It's pretty impressive in this part of the river drainage; we're over a narrow ravine, and the water is some twenty-five meters high. From here the river widens somewhat and then gets narrow again. There aren't too many dwellings in this part of the river as the walls are fairly steep. Downriver though the story changes quite dramatically. The river flows into a wide plain with agriculture land along both sides of the water and several thousand people living in the floodplain, and some live right in the riverbed. We're over them now and are playing a recorded warning about the imminent flooding danger approaching. They don't seem to be taking the warning very seriously; no one is picking up and leaving that we can see.

"We're going to switch to live broadcast now and see if we can't get them out of here. First we're going to set down in an open area near a group of homes and cut the noise from our engines. We're down; starting to broadcast now, people are coming toward the helicopter, and they seem more interested. We've got a Squad member who is fluent in the local dialects and is getting their attention now. We're going to lift up and set back down a couple hundred meters from here." There was pause in the audio. "Okay, we're back down. I think we can cover this entire region of the floodplain in about five to six of these landings."

"Are the people leaving yet?"

"Yes, I think we've gotten through finally. Fortunately there is some upland not very far from here that should be high enough to get everyone to safety. Well, we're about through in this part of the river, so we're going to check where the flood is and then move on downstream to warn the next area."

"Do you see the water yet?" Helene asked.

"No, not yet—wait, we're coming over a crest and there it is, it's moving into the second narrow region. The people we just warned have about thirty minutes before the water reaches them. We're turning around now and heading further downriver."

"We'll be back in touch in few minutes, keep the line open."

"Roger, out."

"Helene, we're getting a message from Florida, switching now," Natalia said as Alex appeared on the screen.

"Kristen, we've reached the last apartment complex that the tornadoes went through, and our worst fears seem to be materializing; we've yet to find a survivor. There are hundreds of dead. Bodies are strewn everywhere. It's bad, really bad, and we are going to need more help in here to take care of the bodies. There is a path of virtually total destruction from here to the water's edge. I've never seen anything like this before in my life. Every single building is quite literally reduced to toothpicks. The force of these storms must have been near record, if not record producing in their strength. We've got some pictures coming through to you now from a camera we've set up. All the scan cams that were in this part of Tampa were destroyed, as you can well imagine, so we've set up some new ones. Are you receiving the pictures yet?"

"Yes we are. You weren't exaggerating one bit; it's looks like a war was fought right through the entire city. Oh, my God, the bodies, they're everywhere," the director stated, slightly overcome at the sight of so many bodies all dead from one tornado. No warning, no reason, just dead. She was used to casualties, but this was getting out of control; thousands dead in Asia and now thousands dead in North America, and it was still happening on both continents. And it would still be days before it was all over. Her concentration returned in less than a minute, and she asked Markus to raise Debra.

"Got her now," Markus said.

"Thanks. Debra, what's your status?"

"We're ready. The winds are picking up along the coast from the Florida panhandle to Houston. Eric is pushing a very large storm surge toward Mobile right now; we're showing at least ten meters and possibly larger. Hurricane-force winds are reaching the coast already, and the rain is starting to really come down. Daria is picking up forward speed and should hit Texas within the next six hours. Hurricane-force winds are reaching the outer islands off the Texas coast already, and gale-force winds are reported one hundred kilometers inland.

Colin is taking aim at New Orleans and should be there in four to six hours. The winds have yet to reach hurricane force in the city, but they're higher than that already on the coast and the tidal surge is pushing up the Mississippi River fifty kilometers. The worst should be in the next six to ten hours across much of the affected region. It's looking as if all three storms will make landfall within a few hours of each other. Right now computer projections are showing that Eric and Colin will probably merge once they hit land, and there is a limited probability that Daria will merge with both of them, and we will have one super storm once inland. The rainfall predictions are anywhere from twenty centimeters to two meters depending on where the storms travel and any high pressure domes that could impede their movement out of the continent. If they stall out for a few days over the same area and have a good moisture flow out of the Gulf, we could see unbelievable precipitation over the central United States and especially the Mississippi River drainage basin. We'll just have to wait and see; however, we already have put out warnings for all low-lying regions everywhere from Kansas to the west, St. Louis to the north and all points in between to the Atlantic Ocean. Also the probability of tornadoes is very high throughout the entire coastal area of the Gulf, and we should start to show some on triple Doppler radar any time now. That's out update here. Anything else you can tell us?"

"You seem pretty current on all the data, but maybe we should consider moving the northern warning line to Chicago; some of the computer projections show the storms reaching that far north."

"That sounds like a good idea, and in just a few minutes we'll be sending the warning out. Anything else?"

"Not at the moment," Director Verba said. "We'll be in touch every half-hour until the worst is over. Good luck to everyone." She quickly switched back to Alex and asked, "What's the strength of the storm now in Florida?"

"It's pretty much over in central Florida, although we're still getting some wind from Eric on the western side of the state, and in the panhandle the wind is really picking up, nearing hurricane force, and the rain is starting to come down in sheets. We're showing some tornado activity near Panama City, but nothing has touched down yet, as far as we know. In southern Florida the sun is back out and the weather is completely calm. What a contrast to the northern part of the state."

"I'll say; how's the rescue effort in Tampa going?"

"Not good at this point. We have eight hundred plus confirmed dead and at least twelve hundred missing. Thousands are injured and the triage centers are treating them as fast as they can. The severely injured who can be moved are being airlifted further south where they can find hospital space, as many as possible are being treated and released, and the most critically injured are getting what hospital space is still available in the local area. We're getting pretty thin on supplies, but we're supposed to have another shipment here within the hour. It's being airlifted from Miami and left minutes ago."

"Roger that, we copy it on the way and some forty-six minutes out of Tampa," Kristen said. "The supplies are on small vans on large transports, so once they land, they can literally drive out the back of the plane and right to the triage centers. This is the fastest way to move gravely needed items. So you should have them shortly. How's your blood supply holding up?"

"We have about one day's worth at the current usage rate, but that usage may go up as we find more victims."

"Looks like the supplies coming in from Miami contain just about everything except blood," Kristen stated after glancing at the inventory of the incoming supplies.

"I thought there was plenty of blood coming on that shipment," Alex said. "What happened?"

"Apparently there was contamination of some sort; we don't have all the details at the moment, but we're working on it. We also have found a supply of blood in Atlanta, and it's on the way. It will be at least six hours before it gets there; hopefully that will be soon enough."

"It should be. Good thing you noticed, before the situation here got desperate."

"I'll check back with you soon, Alex. We're getting an incoming call from Helene. Bye for now."

"Roger, out for now."

"Director, Helene is on the screen now," Natalia announced.

"Helene how's the flood situation?"

"Our warnings are helping move quite a few people to safety; but we're not getting everyone. Some just don't seem to understand the threat. It seems as if they think the storm is over and aren't paying any attention to us. And we're running out of time. The flood surge is moving into the first area we warned right now. We have a chopper there now, and we're switching on the video so you can see what's happening."

"We copy the pictures. Wow! That is some wall of water coming out of the canyon. When it enters that open area, it should really spread out—there it goes. The first buildings are being washed completely away, no trace, nothing left. Everything is going down the river. So far I haven't seen any people; it looks like they believed the warning here any way."

"We're picking up the other helicopters downriver. They are still trying to warn as many as possible. We've projected that the flood surge will become a lessening danger to life and property as it spreads out, finally becoming harmless about one hundred kilometers from right about here. That still puts thousands in jeopardy. We just don't know for sure how many lived here before the storm started, how many had evacuated, or how many had already moved back in after the rains ended."

"What is the computer prediction for the amount of time until the wall is relatively harmless?" the director inquired.

"At the present velocity of about thirty kilometers an hour, just a little over three hours. That is of course without any additional water sources joining this one. There are some other dams along the way that are showing signs of stress, but at this time we are not projecting them to become a problem. Let's keep our fingers crossed though just to be on the safe side. Switching back to the water wall shots …"

"Director," Markus interrupted, "Debra is asking to talk to you—switching her on now."

"Helene, gotta go, will be back in touch real soon. Out for now."

"Director, Debra here. We're really starting to get pounded now along the entire Gulf coast. Winds are above hurricane strength just about everywhere that we are monitoring with our automatic weather stations along the coast. Tidal surges are huge, especially from New Orleans to Brownsville. We've gotten reports now of ten- to twenty-meter tides, and high tide is about three and a half hours away. No reports of any lives lost, but damage reports are already trickling in from trees going down, damage at marinas, buildings being smashed up.

"Also we have our first active line of waterspouts moving off the Gulf into the panhandle of Mississippi," Debra reported. "They should reach land in about thirty minutes at present speeds. Almost everyone has evacuated that part of the state, but we sent out radio and reverse 911 cell phone warnings just to catch any diehards who never left. Another report is just coming in—more tornadoes, this time near Baton Rouge. They're already on the ground, moving northeast at about ten kilometers an hour. Well that's the latest update. I'm sure it will change minute to minute for a while."

"We should start talking every fifteen minutes from now until the worst is over, Debra. I'll call you back in fifteen, but if anything major changes, call us immediately. Good luck to everyone. Bye," the director said, quickly returning to Helene on another screen. "Helene, where's the water now?"

"As you can see from the video cam, the water is spreading out quite a bit, and consequently the water level is receding somewhat, which is good news. The Ganges River is cresting right now, and we should see some improvement there in a few hours. It will take up to forty-eight hours for the water to significantly drop and a bit longer to get below flood stage completely. A lot will depend on what happens. If there is no more rain anywhere in the region for a few days, no more dams breaking, and we keep the sun out to help dry up some of the area, we should be all right in five days at the most. That is the projection for now, although some computer predictions put rain back in the forecast in a couple of days because of a front approaching from Tibet. It could get stalled out, in which case we're in the clear. We'll just have to wait and see."

"The storm in the Gulf—make that the three storms in the Gulf of Mexico are all coming ashore at once, so I need to switch back there for a bit. I'll call you back in about fifteen minutes to see how it's going. Check on your supplies, and see how they're holding up. You can fill us in on any needs you have when I call you back. Out for now."

"Will do, Director, and give them our best in the Gulf," Helene responded.

"I'll do that," she answered and then asked Debra her latest status.

"It is getting serious, very serious, in fact. We've got two category-five hurricanes and one category-three—make that a category-four storm, as it has intensified even further. This has never happened in recorded history before that three storms of such strength have made landfall at the same time like these three are doing right now. Winds are leveling buildings now, tossing boats, knocking trees over. The rain is coming down at the rate of fifteen centimeters per hour in some locations, and flooding is already a problem in most low-lying areas. I think we should send out warnings to any low-lying area:

areas along streams, rivers, and lakes; basements; areas below dams or dikes—if it's not already evacuated, then it needs to be evacuated now."

"I agree," the director stated, "especially after what's happened in India and Bangladesh. We could see some dams give out under this much water. There is still time north of the coastal regions to evacuate the areas that you have described; we'll get the word out right now." With a nod to Natalia, the message was on its way to the millions of North American inhabitants that would soon be affected by the three storms.

Susan was monitoring the weather satellite images from space and processing them through computers to give models of possible storm paths and intensities. She was watching such a model develop and it was showing a high probability for Colin and Eric to merge together when they came onshore and a slight probability for Daria to join them also, after it made landfall. She reported the computer's current conclusions to the director.

"What probability did you say the computer was giving that they would merge, Susan?" Kristen asked.

"Around 83 percent at the moment," Susan answered. "There is a high-pressure ridge to the north that will most likely force both storms toward the east once they strike land, and Colin is the more powerful of the two storms and moving faster, so it should overtake and absorb Eric by as early as tomorrow morning. The combined strength of the two storms could be awesome. The rain potential prediction is astonishing; I've never seen a rain forecast like this before. There is a chance for up to two meters of rain to fall over the next seventy-two hours throughout the southeastern portions and maybe even the north central to northeastern sections of the United States.

"Daria could go either way at this time, according to the computers," Susan stated. "There is a 50 percent chance that it will veer east and in all probability merge with the other two storms and a 50 percent likelihood that it will move to the west of the high pressure dome and could even approach us here in Colorado. Remnants of hurricanes have ended up here before and have caused some flooding in the past. If they set up just right, we can get the classic upslope conditions along the mountains that bring heavy rains all along the front range of the Rockies. We'll just have to keep monitoring the situation for now to see what path it will take. The other two storms are very likely to follow the course that the computer has predicted at this time unless some big change occurs in the current weather pattern, which seems very improbable, but of course when you're dealing with weather not impossible."

"So we could end up getting hit by these storms even here," Kristen said to no one in particular.

"It looks like there's a chance, and in fact, a pretty good chance. I'll let you know when the computer has an update of the model," Susan replied.

"Director, Alex on the screen for you," Markus reported and switched on the audio of the Florida scan cam.

"Go ahead, Alex, what's up?"

"We've run into another apartment complex badly damaged, and our blood usage has increased dramatically; any word as to when that shipment of blood is scheduled to arrive here?"

"We're checking; it looks like it will be longer than first thought. They are being forced to alter their course around the storms, which means that they have to fly out over the Atlantic Ocean then south to Miami, then north to Tampa. So at first they predicted six hours, and they now are saying eight total—so another six hours to go from now. Can you set up any donation areas in the Tampa area?"

"It would take longer than six hours to screen any blood we could get even if we could find donors and spare some personnel to work with it. We're spread pretty thin right now, as most Squad members are busy just digging bodies and survivors out of the huge piles of rubble that were once buildings. We'll run out in about three to four hours at current usage and will just have to get by until it arrives. Maybe we can cut back on some of the marginal uses and save it for the really seriously wounded until the new supplies get here."

"That sounds like a feasible plan for now. Meanwhile we'll scan all closer cities and see if we can't scare up some a little closer. We'll be in touch; signing off for now," Kristen said as Helene came back on the screen.

"Director, we checked on our supplies here, and we're short the following items that we're relaying to your computer right now. The worst shortage is clean water. Almost all the water now is completely contaminated. With so many dead bodies in the water—well it's a dire situation. Also most of the wells have been flooded with contaminated water, so they're all infected. We will become critical within a day, day and a half tops."

"We'll get right on it, Helene," Kristen answered, and Paul was on the task in seconds, juggling computer and telephone. "How's your blood supply?" she asked.

"Adequate at the moment, although we don't have any surplus from current usage predictions. We've seen fewer injured than we projected and, unfortunately, more dead than predicted, which has resulted in less call for blood reserves. I don't know if that trend will continue or not. How is Dhaka faring after that band of tornadoes traveled through there? Do they need blood?"

"They took an incredible hit and many losses," Kristen answered. "Hundreds were killed and thousands were injured when the tornadoes ripped through the city, mostly one part of the city. We got an e-mail from Jireco for supplies a few hours ago; blood was not on the list, and like you, water was the number-one item requested. We located some for them and enough pills to make drinkable water for at least a week for the entire population of southern Bangladesh. So we'll be scaring up some more decontamination pills for them too. We've got manufacturers working overtime to get some more pills ready as soon as possible. We'll get you some too as soon as they're ready."

"That should do the trick; what's the timetable for getting them here, a day, a week, what should we be planning?"

"About three days, four max. The first shipment will be airlifted into Asia from the U.S., and more will follow every day, as it's ready. Anything else you need now?"

"Not at the moment, but we'll be in touch if anything changes. Out for now."

"Talk to you soon. Out for now."

"Director, it's Debra on screen three," Susan said, indicating the middle screen in the large control room.

"Debra, what's up?"

"Tornadoes; we've got five different bands moving through the south as we speak. Two in Mississippi, one in Alabama, one in Texas and one in the panhandle of Florida. They are ripping up things, according to reports we're receiving here. It seems that they all appeared at about the same time. We don't have any accurate numbers on the twisters, how many there are in each band and so forth. We do have a pretty good idea what direction they're headed, though. They all seem to be trending to the northwest. It appears that conditions are very ripe for further tornadic development. We should expect many such outbreaks over the next few hours.

"Here's the update on the storms themselves," Debra went on: "winds are reaching three hundred kilometers per hour from Houston to Biloxi; damage is very severe, and the rain is falling at over ten centimeters every fifteen minutes. That's a lot of water. Flooding is extensive, and it's only going to get more intense before it lets up. What can you see from space?"

"Well, the first thing you notice is how big these storms are," Kristen stated. "They cover the entire Gulf and extend well into Mexico, Texas, up into the southeastern U.S., and back into Florida. Then the three eyes are very distinctive although it appears now that Eric and Colin will join together soon and merge into one storm. That will happen within about twenty-four hours, according to the latest computer projection."

"What is their most likely direction? Is it still to the northeast?" Debra asked.

"It appears so at this time," Kristen replied. "Although we don't have an exact course, we do have the general direction that the storms are expected to travel, and it is the northeast. That is the most common direction that hurricanes take after coming ashore from the Gulf of Mexico and moving into North America."

"In a way that would be best, as it would lead the storms to the open water the soonest and get them away from land. Well I'll call you back if anything changes; otherwise I'll talk to you at our next regularly scheduled time. Out for now."

"Roger that."

"Director, Alex on the screen," Natalia interjected.

"Alex, go ahead."

"Director, we've got a real problem we just came across. We've found a school that was being used as a shelter that has collapsed under the stress of the rain and wind. Reports have at least three thousand people in the building. We have talked to some of the survivors that escaped just before the building went down. They said there were loud noises before the crash, and people were trying to get out before the structure collapsed, but they don't think many did get out, as it went down pretty quick after the first sounds were heard. We've got equipment on the way to start digging into the rubble. There is a basement that quite a few of the people were in that may still have survivors. We'll have to be very careful not to push any debris down on them.

A report is coming from some of the Squad members near the building," Alex continued, "and they are relating that they hear voices from the lower portion of the building. We are digging down beside the school now, trying to reach the basement level without going through the building itself. We'll punch a hole in the wall as soon as we get down a couple of meters or so. We're really going to need that blood here now. Our supplies will be completely depleted after we start getting victims out of this building plus all of the other injured that we're coming across throughout the city. What's the timetable for that blood shipment now?"

"It's still on schedule; should be there within two more hours. They were able to shave a some hours off the trip, as the storms have moved to the northwest from southern Florida, so they don't have to fly so far south before coming your way. We're looking around the rest of Florida also, to see if any other blood sources could be tapped, but so far none that anyone can spare. We'll keep you informed of the situation. I'll talk to you in ten minutes, Alex. Helene is calling from India. Out for now."

"Roger."

"Helene, how's it going?"

"We're looking pretty good except for the wall of water from that last dam to break. The Ganges is subsiding now and is down an entire meter from the crest several hours ago. But the real problem is this other river that's out of control. We tried to get people out of the way, but they didn't seem to believe us until it was too late. By then the water just washed them away. I'm afraid that we are going to have major casualties along the river, and it is almost total destruction in many areas. The death toll may well reach into tens of thousands plus the loss of hundreds of thousands of domesticated animals and millions of square kilometers of farmland. The aftermath of this storm and the consequent floods will be felt for years, if not decades. For one thing the farmland is under meters of mud, debris, dead animals, and so forth. We're going to need clean drinking water here right away; almost all the available water supplies, wells, and even springs are contaminated. I know that you have supplies heading here as soon as possible, but if there is anything that you can do to speed it up … well, that would great." The accumulating fatigue of the last week and more was evident in Helene's speech and her stifled yawns here and there. It had been days since she or anyone else in the Squad had had any appreciable sleep, and the exhaustion was beginning to show. But it would be still days more until they could sleep soundly. And it would be even longer before life returned to some sense of normalcy.

"We'll see what we can do to accelerate the process, but I'm afraid that everyone is going as fast as they can already."

"I know, but thanks for trying."

"I'll be back in touch soon; keep us posted if anything changes," Kristen said and then switched back to Alex who was coming over the air with a new sense of urgency. "Alex, what's up?"

"Director, we've just picked up on triple Doppler a new squall line with waterspouts moving toward Panama City in the panhandle. It's coming out of the Gulf and is part of hurricane Eric. It's big; although not quite as large as the one that hit Tampa, it's still massive enough to do severe damage if it stays on its present course. Right now we show it moving onshore in thirty minutes and striking Panama City and the surrounding communities of Parker, Springfield, Callaway, Lynn Haven, Cedar Grove, and Bayou George. The edge of the squall line, which is measuring about twenty kilometers wide, could nip as far west as Panama City Beach, but the main portion of the tornadoes will slam right into the cities we just listed. Of course, we've sent out warnings to the entire area, but after what we saw in Tampa, I'm not sure how much good that really does against storm lines like this one. We show between thirty and forty individual waterspouts in the band at any moment. They keep forming and dissipating like the ones that hit Tampa.

"By the way, we're just about through the city now and have found most of the buildings that were destroyed and have commenced rescue efforts on the majority of them already. With some of the buildings that completely collapsed, we're having a little difficulty getting the necessary machinery to the sites to remove debris due to the amount of rubble in the streets blocking passage. We're just running a bulldozer ahead of the equipment now to clear some kind of path.

"We anticipate getting to all of the buildings within another couple of hours. From there it will take several days, at least, to get through all the collapsed buildings. Some of them were up to thirty-five stories, and we have to go slow in case of survivors, which we are optimistic about in some of the structures.

"Here's an update on the twisters ... They're fifteen minutes from land, course still the same ... Thirty-five confirmed tornadoes, another fifteen possible, and they're running quite close together, so it appears that destruction will be almost complete in the affected areas. We've got scan cams in place, although they won't last long once the band hits; switching them on main viewer now."

With a click the picture showed Panama City under the fury of the hurricane already, with more intense winds about to hit. The camera was bouncing around, both demonstrating and showing the horizontal rain and trees leaning almost to the ground under the wind's pressure. They all watched in silence as the time slipped away, quickly bringing the tornadoes ashore.

"Doppler radar is showing the line only a kilometer off shore now. We should be able to see them shortly," Alex said, and then: "There they are!"

The scan cam showed a band of waterspouts throwing up water a hundred meters into the air ahead of them racing toward the coast. Just minutes later, although it seemed a lot longer, they came ashore and in an instant became tornadoes. The wall of twisters swiftly started smashing everything in its path. The control room sat in complete silence as the destruction unfolded before them on the screen, buildings exploding as the lower pressure of the storms swooped down on them. The winds, then catching the flying debris and hurling it upward into oblivion, made for an eerie sight. A second later the scan cam, shaking to the point that the picture became completely blurred, went blank.

"We're switching to another camera further from the shore," Alex interjected into the silence. "It's coming right about—now."

"That was something else!" Paul said. "It's always incredible the power of tornadoes. If they weren't so destructive, they would almost be beautiful in their own way."

"I know what you mean. They are awesome in their own way," Kristen said. Then, at a nod from Susan, she went on, "Alex, we've got a call from Debra and need to go for a moment. We'll be back in touch shortly. Bye."

"Roger that, out for now."

"Debra, how's it going?"

"Hello, Director. We've got flooding starting all over the place. The rain is reaching record amounts per hour from Houston to Mobile and north well into the southern United States. Rivers and streams are reaching flood stages throughout the south central states. And with this storm surge reaching the Mississippi River—well, you can see our projections feeding into your computers. This is based on instruments placed on every major river and stream transmitting the information to a central computer, which then projects how high the main collector rivers will rise. At the present rate of rainfall the Mississippi will reach flood stage in less than three hours and will go some twenty to thirty meters over flood stage within twenty-four hours. It's a good thing that we did such extensive evacuations before this all began; I just hope it was enough. We could reach record flood stage in a few days if this continues for very long. Every possible warning should go out to anyone even close to a river to be alert and watch water levels very, very closely.

"Also we've had over a hundred twisters reported besides the ones hitting Panama City right now. Most are isolated, and many have missed any populated areas. A few have hit some cities, especially in central Mississippi and Arkansas, with several also in eastern Texas and western Louisiana. We've even shown some into Tennessee. The number of confirmed dead is twelve, but communication is out in many areas, and it's too dangerous to be outside in this kind of weather. I'm afraid the numbers will go higher as the storm passes and communication is restored.

"Another problem we've had in the northern part of the affected area has been hail in some thunderstorms. Some of the hail has been grapefruit size and has caused considerable damage. And of course there is the wind. It's still pounding the entire region with hurricane strength or more in eleven states right now, and we could add another two or three in the next few hours. It's going to be pure hell for the next twenty-four to forty-eight hours. Eric and Colin have practically converged into one storm now, covering a vast area. The high pressure to the north is still stalling the storms from heading that way, although the storms have pushed the high-pressure ridge back a little. Still, we are projecting a move to the east, so the Carolinas and Virginia will get nailed. Mostly rain for those areas as the winds are projected to subside over the land as the storms push inland."

"Debra, are there any projections on rain for the East Coast?"

"Just preliminary; we don't have enough data on what the precipitation rate will be by then. One thing is for sure though: if it stays this high, there will be major flooding everywhere."

"We'd better get the maximum warnings to these new flood areas that might be affected," the director stated, and in seconds the word was out.

"Director, word is coming in from Helene," Elawa said, relaying the message from India. "Another earthen dam has given away on a new river feeding into the Ganges. It will crest in three hours and then flow into the Ganges. She went to the scene as it was close to where she is and said that she will be back in touch soon."

"Thanks, Elawa. Patch her in when she comes back on, please. Paul, how's the blood transfer to Florida?"

"Well, let's see … about an hour and forty minutes has passed since we last checked, so that would put it about twenty minutes out of Tampa and should be to the needed sites in about thirty minutes after that with any luck."

"That's good. Feed that information to Alex as soon as you can."

"Will do, Director," Paul said, and then added, "Also, just so you know, we have found another couple of earthmovers and have them on the way to India. We're still searching for any other equipment that we can requisition for a little while to help assist the rescue efforts under way."

"Good work. Let's hope we won't need too much more equipment and that this thing is over soon," Kristen exclaimed.

"Director, this is Debra, and we're showing the rain actually increasing if that's possible. It is reaching record levels now and we're in new territory for rainfall. The current flood projections are staggering. We'd better insist on new evacuations in many areas that were not covered in previous flood warnings, and in fact, all areas anywhere near water need to be cleared out now."

"We read you loud and clear, Debra, and we have sent the word on the way already," Kristen replied.

"Director, it's Helene …"

"Helene, are you alright?"

"Yeah, I'm fine, and so are all the Squad members right around here anyway, but we have a major problem with this new dam breaking. The river that is downstream next is on the verge of a major change in direction, and it looks like this will be enough water to push the river into its new path. There's not enough time to enhance the banks to a sufficient level to stop the river from changing course. Our prediction is that the water will be to the point in question in less than three hours, and we don't have any equipment even close right now. So I think our best bet is to evacuate the new river course ASAP. We've sent the word out, but as you know we've had some real trouble getting people to believe that any more can go wrong and especially with the storm pretty much over. The rain is all but stopped, and the winds are back to their normal strength along the coastal regions and only slightly higher inland than normal. The sun is out in most of the region that was hit by the storm. Any ideas?"

"Where is the closest piece of equipment that is capable of building the banks up?" the director asked.

"We checked just a few minutes ago, and they are about four hours away by land and under an hour by air."

"Where's your closest helicopter capable of lifting that type of equipment?"

"Let me check—looks like about an hour from the equipment and I suppose the next question is how long will it take to repair the banks once they get there, right?"

"You got it and we'll have the answer in just a moment"

"While you feed that into the computer, how's North America faring?"

"The south-central and southeastern United States are being pounded right as we speak. Tornadoes are ripping through cities everywhere and the rain is at record levels throughout the area. The flooding is going to be severe and property damage extensive. We're hoping at least that the death toll will be low as there were extensive evacuations of all low-lying areas before this thing got started."

"Director, we've got the answer," Natalia said. "It will take about two and a half to three and a half hours. We already got the project under way at the first mention of this so as not to waste any time, and the helicopters are already in route toward the earthmovers. They should be there in about fifty minutes. We've relayed the urgency of the situation and they are going to push the throttle to the max and see if they can't shave a few minutes off that flying time. The earthmovers are being moved into position with steel cables attached so the helicopters can just move in and pick them up. We also are starting evacuations in the possible new river path. The choppers just reported that they are ahead of schedule ... good news. The drivers of the equipment have reported that they will travel in the cabs of their earthmovers so that they can start driving the instant the equipment touches down and the cables are released. That ought to be heck of a flight!"

"No kidding," Kristen replied. "Good luck to them, and we'll check back in as they close in on their destination or when anything else happens. Out for now, Helene." She immediately shifted back to Debra in the United States. "Debra, do you copy?"

"Yes, Director, we're here. Have you seen the new rain predictions? It looks like they will pass all previous records by leaps and bounds. This is going to get out of control rapidly. I would suggest total evacuation of all areas under twenty meters of flood stage immediately. Plus the likelihood of dams breaking, levees collapsing, and lakes overflowing is extremely high. We're definitely in for some extreme times throughout the southern United States, and there's not much we can do about it, except get out of the flood areas. One piece of good news at the moment is that the winds have dropped a little. That's our update here: rain, rain, and more rain!"

"We'd better get moving on the evacuations," Kristen replied. "We'll contact every governor in the affected area and get them to declare mandatory evacuations for the regions getting drenched." She then turned to Natalia and instructed her to get the governors on the air. As each of them came on the screen, the director explained the situation and what needed to be done to save as many lives as possible. The governors received the warnings soberly and went right to work to issue the appropriate orders. This would now be a race against the clock. The waters were rising rapidly, and even though millions had already moved to higher ground, it just wasn't enough under the circumstances now prevailing. "Markus, can you raise Debra again, please?" she asked.

"Here she is now Director."

"Debra, we just finished talking to all the governors in the southern United States, and they have sent the word out to get people moving now. The problem, of course, is that communications are down throughout most of the region now, so it will be hard to contact anyone, plus the roads are nearly impassable everywhere. If they can get the word to the people, there will still be close to insurmountable problems in getting people to higher ground. Most people won't even know where higher ground is, let alone know how to reach it while dealing with flooded roads, washed-out bridges, and most road signs gone. This evacuation will be very challenging. We can only hope that there are enough people with battery-operated radios to hear the warnings and then spread the word around their local area."

"You can say that again," Debra answered. "We'll do what we can from here to assist, but all our planes are grounded, boats are in dock or out of the water altogether, and most ground vehicles are stranded throughout the south with no roads to travel. What else can you tell us?"

"Not much at this time that you don't already know. But I'll tell you this: we're going to have an enormous amount of work to do when this is over. The cleanup alone will take months to a year, and the healing will take years longer to get over the losses that have occurred from these storms."

"That's for sure, and—Wait, another tornado has just shown up ... moving northeast of Jackson, Mississippi, toward Sandhill, and still another one is developing just southwest of Leesburg and moving toward the center of the city. They are moving fast and we are barely going to be able to get a warning out ahead of them. The tornadoes just dropped out of the sky, no warning, no indication whatsoever that they were going to appear. There have been so many of this type of twisters in these hurricanes, just dropping out of nowhere and wreaking their destruction on unsuspecting people. The tornadoes have caused a great deal of the destruction that these storms have delivered, second only to the flooding that is occurring everywhere.

"Speaking of the flooding," Debra went on, "some areas are now some fifteen meters under water, and it's still rising. We may see regions as much as thirty, thirty-five, maybe even forty meters under water before this over."

For a couple of minutes there were no reports from anyone, just the pictures on the screens from the scan cams that were still working. Over half of all the scan cams in the affected region were out; of the ones that were still working, many of those were shaking or flickering, blinking on and off in the winds still ripping through. The eerie pictures showed the mostly horizontal rain coming in sheets; sometimes the pictures went completely gray with no detail from the infrared enhanced cameras. The night was fading, and the early morning glow from the east was barely visible through the dark clouds delivering their load of moisture to the saturated ground below.

The silence was broken with a call coming in from Helene in India. "Director, do you copy?" she asked as her image appeared on the main screen in the control room.

"Go ahead, Helene, we copy you loud and clear."

"Kristen, just to let you know, the earthmovers are in place and have started the repair of the banks. Night is falling here and we will be working in the dark for the most part. We do have some floodlights in place, and the machines do have some pretty good lights on them, so we're confident that we can get the job done before the river overflows its banks. Here is some footage of the choppers bringing in the earthmovers." At her words the pictures starting coming over the screen. It was an amazing scene: giant earthmovers with their lights on swinging in the air under helicopters just as big.

"Here are the machines landing, and as you can see, the earthmovers had turned on their engines and hit the ground running. They started hauling dirt from below the banks and moved it to the lowest areas first. We are anticipating that the water will be rising near flood stage before we've finished, so we will be playing damage control and hit the areas that are about to flood next," Helene said as the pictures continued to roll.

"And here is a live picture. As you can see we are still ahead of the rising water, though not by much."

"It looks like you'll be successful if the water continues to rise at its current rate. Let's hope so anyway. Well keep us informed, and we'll check back in with you in half an hour," the director said. With that, she walked back to her office with Silverton right at her heels. The cat jumped on her lap when she sat down and began purring and rubbing against her hand that was lying on the desk. She continued moving back and forth for several minutes while Kristen scanned her e-mail.

There was a message from Kristen's mother thanking her for the flowers that she had sent to her uncle's memorial, which had taken place earlier today, a message from Karen asking how she was doing with all that was going on in the world, and a brief note from David. He was back in Colorado. And he realized that she was most likely busy with all the storms going on, but would she be available for dinner after the worst was over and she had a free moment again? She smiled her first smile in days. David was back; he was always so much fun, so full of life, and just what she was going to need after all this was over. She laid her head on the desk and just for a moment fell asleep.

CHAPTER FOURTEEN

Kristen Verba woke up with a start. She hadn't meant to fall asleep; it had just happened. She had been completely exhausted. Her calico cat was sleeping while purring away. Now as she glanced at the clock and jumped up to race back to control headquarters, she realized that she had been asleep for only twenty-three minutes, but it was enough to make her feel somewhat rested, in some strange way. It was morning now in the Western Hemisphere and night in India as Helene and all the rest of the Squad members there were hard at work on their jobs. Kristen ran down the hall back to the control room and entered, glancing at the screen. Silverton ran in and jumped up on her favorite computer to survey the room.

Debra's face filled the screen as Kristen watched. "This is Debra. Do you copy?"

"We do," Kristen replied, walking toward the screen. "What can we do for you?"

"Nothing at the moment, but we are going to need a lot of assistance as soon as this rain stops. We're reaching flood levels along almost every river, stream, and lake south of the Mason-Dixon Line right now. And the worst news is that the rain is still falling, and in some places falling heavier now than before. This is getting completely out of control," she said. You do all you can and it is still not enough. She was experiencing this now.

"Kristen, there is one thing that you can do for us. Water. We're going to need drinking water, as every normal supply is going to be contaminated."

"Director," Indira spoke up, "we were anticipating that and have already set up water shipments to be made just as soon as possible after the rain stops or whenever it's safe to fly again."

"Good thinking. Quite a few of the airports will probably be underwater though, so where are you thinking of dropping the water?"

"Well, most of the water will be coming in on helicopters," Indira continued, "and then we're linking up with small craft that can distribute the water in the flooded areas. In areas where we can land, we're sending in transport planes. Also we are sending ships loaded with supplies, including water, to the port cities. But that will take some extra time so we will drop in emergency water by air first."

"Sounds like a plan. Keep us all informed on how it is going."

"Will do," she replied.

"Debra, did you catch that?"

"Yes, most of it—enough to get the drift of it. It's good that water will be on the way shortly. Did I hear that some might come by water?"

"Yes you did. We could get some to New Orleans, Mobile, Biloxi, Houston, Pensacola, and other ports. It would be in tankers and could be pumped ashore anywhere they would like it," Kristen answered.

"Well that could prove to be really valuable in those coastal cities, because they were some of the hardest hit areas and will need water the most. Are the ships on the way?"

"Yes they are. They are leaving from South America as we speak. The water in the southern Caribbean is calm enough to sail on already, and as we move north the water will be calming down as we go, at least enough to keep right on going. Or at least it appears that will be the case under the present conditions," the director responded.

"Director, we just received a message from the water ships, and they are exactly on course and on time," Indira told her.

"Great to hear. You've got to be happy over any goods news in situations like this," Kristen said.

"That's for sure, always be thankful for what you got," Debra stated as a matter of fact.

"Helene is on screen one, Director," Markus said and flipped the control to put screen one on the big screen. He glanced over at Natalia and they smiled at each other. It had been days since they had been together. They yearned for each other in a big way, but they had a job to do, and they would have to wait until this was all over before they could be together again. Their smiles to each other were enough for now.

Helene began, "Director, here's the latest update on the river repair. It looks like we made it. We have reports that the water rose within just a few centimeters of the top of the bank before we raised the embankment high enough to stop the water. Now the water is remaining at the same level and we're raising the bank quite quickly now above flood stage. We are waiting for final confirmation from Squad members at the site, and I'll let you know as soon as we hear something."

"Great to hear. Do any other areas need immediate attention?"

"Not now. We're finally starting to get this place under control. I think the worst is over, cross fingers and touch wood," she said giving the old Australian superstitious saying.

"The satellite picture would agree with you," Kristen said. "The sky is clear in all directions now, and the forecast is for continued good weather at least for a couple of days."

"Well this storm is over. Now we just need to end the ones in the Western Hemisphere and we can take a breather again," Helene added.

"And boy do we need it; we're near exhaustion here at headquarters. I think the group here has been awake for nearly four days straight. We've had almost no breaks since this all started and it is beginning to show," Director Verba replied.

"Director, it's Debra on the line," Natalia inserted.

"Helene, Debra is calling; I'll be back with you in a moment."

"Okay, Kristen," Helene responded.

"Debra, we copy you. What's up?" Kristen asked.

"We are getting new reports in from the field, and we've got rivers overflowing everywhere. Five dams have given way within the last hour in Alabama, Mississippi, Louisiana, and two in Georgia. They were small dams, but it appears that it may be only the prelude to what's coming. We might end up with some large dams breaking before this over." Debra went on, "Also we've had another twelve tornadoes in the last half-hour touch down all across the southern United States wreaking havoc wherever they meet the ground."

"Where did the tornadoes touch down?"

"Let me see here—two in Texas near Houston, three in Arkansas, one in Mississippi, two in Alabama, one in Georgia, two in the panhandle of Florida, and one in South Carolina. The worst damage has been reported in the Texas twisters. They hit a shopping center and apartment complex and inflicted major destruction on both."

"Any word on casualties?" Kristen asked.

"Not yet. We've been trying to reach Squad members in the affected areas, but communication is difficult everywhere in the region. Even radio contact is limited right now as some areas are under water. And since the tornadoes, we've not heard from several more of our people. I'm sure they're trying to reach us, so we should hear from them soon if they are able to communicate."

"That would be normal procedure, so you know they will as soon as they can."

Debra said, "We do have some reports from other areas. Let's see … the tornado in Georgia went through some farmland and destroyed several houses, barns, and other smaller structures. In South Carolina the information we are getting is six people dead and twenty-eight injured as the twister ripped through a shelter for people who had to evacuate their homes from flooded regions along the coast. How ironic, how sad." There was sadness in her voice that she was trying to cover, but doing a poor job. "They fled a dangerous situation only to die in another. That's life."

"That's for sure. We all do the best we can to survive, but then something stupid comes along and rips the soul right out of our existence. It's so random, so uncaring, so ..." Kristen trailed off, her thoughts with the unfortunate victims of still another twister.

"Director, Helene is calling for you," Natalia said, and then switched Debra to the small screen and Helene to the large one, after informing Debra that Helene needed to talk to Kristen.

"Helene, what can we do for you?"

"Just wanted you all to know that we have confirmed with sources on hand that the river did stay within its existing banks and that the repair job was successful."

"That's fantastic. It's great to get a little good news around here; it's pretty bad right now in the southern United States, tornadoes ripping through many cities and mounting casualties. We are experiencing the worst weather in the modern history of North America, and it's only going to get worse before it gets better. We've got about another thirty-six hours of real intense weather, and then the floodwaters will still be a problem for days after that."

"Well, I think you can now downgrade the situation here from a level five to level three," Helene said, "and hopefully back to level two by this time tomorrow. The most critical situations are over. Sure we still have some flooding and water shortages, but all the problems are in an improving mode. Speaking of water, how does it look to get more drinking water here?"

"Let's see." Kristen turned to Paul, who had been working on the procurement of clean water.

"The first shipments should arrive there by the morning," Paul responded after checking his computer, "with regular deliveries every day after that as long as they need them."

"More good news, that's great," Helene answered back.

"Helene, we'll check back in with you every six hours, or if something comes up, call us. I think you all deserve some rest now to be fresh for the rebuilding task ahead. Take care, and thank everyone for how well they minimized casualties and damage, considering the magnitude of this storm. Out for now." The director motioned for the screen to switch back to North America. The cat woke up and ran over to her legs and started rubbing up against them. Kristen unconsciously picked up Silverton and put the cat on her shoulder. The calico was, of course, purring.

"At least half of these two storms have finally come to an end," Natalia said to no one in particular. "We just need the strength to endure the next couple of days and I think we'll be through the worst.

"Markus, what is the latest position of the storms?" Kristen asked.

"Well, let's see, Eric and Colin have merged over land, and Daria is still heading into Texas. The center of Colin-Eric is near the city of Tuscaloosa, and the steering winds aloft are right now from the west-southwest to the east-northeast. Although the steering winds are almost stagnant, they are still moving the storm mostly east now and just slightly north.

"The bad news," Markus continued, "is that because the steering winds are so weak, the storms are slowing down, which, as you all know, means more rain. Triple Doppler radar has rainfall amounts approaching the one-meter mark throughout the Southeast U.S., and it's still just pouring. At this time every major body of water that we are monitoring is in flood stage. That has never happened before as far back as records have been kept. This is truly an impressive storm."

"That it is," Kristen replied, awestruck at the report.

"Director," Debra said, "we've just received a report from near Little Rock of a major levee giving way, and an entire city has just been washed away! We don't have any more than that right now, but we got Squad members in the area and we should have more news momentarily. Wait......, we're getting a report, we'll put it on audio"

"Debra, this is Alice near the levee … we're above where the city is located or I should say used to be….. it's completely gone! Nothing! Nothing! Nothing left at all. The rain is increasing, but the winds are decreasing. The real problem is water, way too much of it. This might be only the beginning of these problems; we're seeing waters reaching the top of dams, levees, and riverbanks everywhere. If the rain continues at this rate, the entire southern United States could be in some form of flooding jeopardy. We're going to suffer some severe casualties at the present rate of precipitation."

"Alice, this is Director Verba of the Squad. Can you hear me?"

"Yes, I can, what can I do for you?"

"How many people were in the city when the water hit?"

"We don't have an actual count, but the estimate was near twelve thousand to thirteen thousand."

"Had we posted any warnings to these people?"

"Yes, we've given warnings to everyone in any location that might prove hazardous, but most people are choosing to ignore the warnings or can't do much about it even if they want to because so many roads are out, bridges gone, and so forth. We can put out new urgent messages citing this as an example, but frankly where can a lot of these people go? Everywhere is in danger from flooding except the tops of hills or mountains, but how do people get there?"

"We can only do the best we can considering the circumstances," Kristen answered. "We should put out the warning, but acknowledge that the flood danger is extreme everywhere and that everyone should seek the highest ground possible as quickly as they can. If there is any way to get people completely out of the region, via helicopters or some other way, then that should happen. And we should take whatever precautions we can to get people out of low-lying areas for sure."

"We'll do the best we can," Debra said, "but the rain is falling so hard in most regions that it's almost impossible to get any aircraft into the air. What we really need are boats; it may be the only way to get people out, just float them to safety."

"Maybe that is the answer," the Director replied: "boats. Boats of all types; we'll see what we can get our hands on and get them to the people who need them. It may be the only answer, as there may not be any dry land waiting for the inhabitants, even if they do get our warnings. Boats could be dropped into the areas with the least chance of having high land, and at least it would give them a fighting chance."

She then turned to Paul, who nodded back and immediately went to work finding boats that could be easily transported to the affected areas. The boats would need to be near the areas where the flood danger was the most extreme, as there would be no way to fly or truck them anywhere in view of the weather. The rain kept falling, and the urgency grew by the minute.

"Director," Paul reported back in just minutes, "we've located quite a few boats and have arranged for them to basically be floated into position."

"Good work. Do we have any windows in the storm coming?" Kristen asked Indira, who was working the weather monitoring station.

"We've got pretty tight squall lines for the next couple of hours," Indira responded, "but then we will have about a ninety-minute opening before the next squall line moves into central Arkansas, which is the most affected area at the moment."

"Great, let's be ready to take advantage of that break in the weather and get the boats where they need to be," Kristen said. As the break in the weather came, Squad members were busy delivering, by whatever means was available, every boat they had mustered up.

Even though the rain was still falling, the wind was dying down enough for the people to escape in the boats where at least they would be somewhat safe from drowning. When all was said and done, almost three thousand boats that could hold in excess of forty five thousand people had been put into use. And they would need them, as the rain began to come down hard again.

"Director, we were successful with our mission," Debra reported. "Already many people are escaping in the boats, and although they aren't reaching dry land yet, they are, at least, in a safer position as long as the wind stays down from the record levels earlier, and reports coming in from the region show the wind subsiding across the entire storm area with the exception of eastern Texas where winds are holding steady. However, the computer projections indicate that the wind will be dying down everywhere soon as the storms move further inland.

"The bad news is that the same projections show the rain increasing pretty much throughout the same area for at least the next twenty-four hours before the precipitation tapers off. Then we should start to see the beginning of the end of this thing."

"Well that's good news. Light at the end of the tunnel. Then the cleanup begins. But that's a whole other story. Debra, what's the known casualty count?" Kristen asked.

"We only show 374 confirmed dead; the unofficial count, though, is well over three thousand and could soar when we get into that city in Arkansas that is missing. We have people on the way, but the going is slow. The helicopters had to leave the area when the rain made it too dangerous to fly. We are showing a break in the squall lines in about an hour and will get the choppers back in the air there. The ground approach will have Squad members in the region in about two hours. They are traveling across the tops of the ridges standing clear in the water-swollen valleys. We've been in touch with them as they approach the disaster area—though you'd have to call the entire southern United States a disaster area."

"That's for sure. Well, keep us informed when you hear anything new. Talk to you soon. Out for now," Kristen stated.

"Director, it's Alex on the screen," Susan reported as Debra signed off.

"Thanks. Alex, do you copy us?"

"Loud and clear. I just wanted to give you an update of the Florida situation. First, except for the panhandle region of Florida, the skies are pretty much clear and the storm is now completely over. We have relief efforts under way across 98 percent of the state. Again everywhere, except in scattered areas around Panama City where the tornadoes hit, we have units in place. We should have Squad members in all affected regions within the hour. We have treated nearly half a million people for injuries: the majority of them for cuts and bruises from being hit by flying debris or collapsed buildings and for dehydration. The death toll stands at 3,947, and we're still missing quite a few thousand; we expect to find that most of them are in shelters or have fled their homes and are all right. We are tracking down missing people constantly right now and hope to find the majority before the day is over.

"We lost 132 bridges, and seven major airports are still closed due to damage. Thousands of public buildings, tens of thousands of businesses, and nearly three million private residences have sustained some level of destruction, from broken windows and damaged roofs to being completely flattened.

"Every waterway in Florida flooded at some level," Alex went on, "and we had numerous levees break. Flooding caused extensive damage everywhere, but wind and tornadoes caused the worst destruction. The forested areas were badly hit, especially in central Florida, but the northern part of the state and the Everglades suffered only minor damage, which is very good news for the National Park. More good news is that the Park was suffering from a two-year drought, which was ended by over a meter and a half of rain that fell during the storm."

"Well, it sounds like things are under control in Florida," Kristen said, "at least in central and southern Florida. Is there anything else we can do for you or that you need for the relief effort?"

"Not at the moment. We received our water supplies, medical supplies are in good shape, and food and shelter are adequate for the time being. We will need to stockpile rebuilding materials, especially for the infrastructure that's been destroyed, as soon as all the missing are accounted for and the injured are treated. We're starting to put together a list of just what will be needed, and we'll send that to headquarters as soon as we have it ready."

"We'll get to work on your list as soon as we get it. If there's nothing else for now, I'll talk to later," the director said.

"Sounds good, over and out."

"Natalia, have we heard from Helene lately?"

"No, we haven't Director. Do you want me to raise her?"

"Please."

"Stand by; I'm calling her now … Switching her to the big screen."

"Helene, do you copy me?"

"Loud and clear. What's up?"

"How's it going? What is your current status?"

"Well, it's going pretty good here considering everything. We were able to keep that one river within its banks and avoid a major disaster. We weren't so lucky on some other dams breaking, and I'm afraid the casualties are running fairly high. We estimate nearly two hundred thousand lost, and there are still a million missing at least. But we do think most of those people are just displaced and not dead. We also have treated four million injured, most with only minor injuries, but we are reaching full capacity of available hospital beds. We're hopeful that we have dealt with most of injured by now and the need for further hospital space will dwindle."

"How are your supplies holding up?" Kristen asked her.

"We just sent a request in for more water and blood supplies. Otherwise we're doing pretty well. Our food reserves are looking good, medical supplies are holding up, and shelter is not much of an issue right now."

"Then you'd better start your list of what you'll need to begin repair of the infrastructure destroyed."

"We've already got teams on it," Helene said, "and we expect to get a preliminary list to you within twenty-four hours. It will be a week or more before we survey the entire subcontinent and completely assess their needs. Bangladesh was hit hard with tornadoes, and they will need a lot of help. India had more problems with flooding, as did Nepal where the storm finally rained itself out. Southern China's main problems were also due to flooding. The storm had mostly rained itself out before it reached China, but the southern waterways were still hit hard.

"That is pretty much the report from here. How's it going in North America?"

"Not good," the director answered. "We're reaching record rainfalls all across the region, and flooding is out of control with every pond, lake, stream, and river over its banks. We've had a major dam break washing away an entire town of about twelve to thirteen thousand. And we've got at least another twenty-four to thirty-six hours before the rain starts to let up. Then it will be days before the water levels begin to decrease. After that it will be just like the situation in Asia; months and months of repair, rebuilding, and just starting over for many people."

"Yeah, it'll be years before everything is back to normal, but now we need to get all the injured dealt with and start the cleanup. I'll call you back in a few hours," Helene answered.

"Okay, and you should see that everybody in the Squad gets some rest there pretty quick," Kristen suggested as Silverton began to rub across the front of her legs.

"That sounds like a good idea. We're all beat, as I'm sure all of you are by now too."

"That's for sure."

"Director, it's Debra; I'm switching her to the big screen," Markus said as he brought up her image.

"Debra, do you copy?"

"Loud and clear, we've got more problems.… is there any way we can secure more boats? Several more areas are going to be completely underwater within two to three hours at the present rate of precipitation."

"Where do you need them?" Kristen asked.

"One place is near Greenville, Mississippi," Debra responded, "and the other just north of Dumas, Arkansas. Both are near where Arkansas River flows into the Mississippi River, which is over its banks from St. Louis to the Gulf of Mexico, and the further south you go the more severe the flooding. In fact, we need help at several points up and down the Mississippi. The rain is still coming down at about the same rate as over the last five hours; but good news is that the wind is still decreasing across most of the region, and—knock on wood—we haven't had any tornadoes for several hours now. We may have passed the peak with the wind and tornadoes, if only the rain would stop."

"What if we send larger boats up and down the Mississippi?" the director asked, thinking out loud. "Have people get as close to the river as they can, and we'll pick them up with the life boats on those ships."

"That should work," Natalia said, picking up on the director's idea, "although with the amount of water coming downstream, we will be better off to have most of the boats travel with the flow." She continued, "And if the wind lets up a little more, we could have helicopters pick up a lot of people pretty quick and move them to safety."

"Find out where any steamers, barges, and ships are," Kristen directed, "anything that could hold people are, and get them in motion where it's safe to do so and get some helicopters to work with them." In moments everyone in the control room was busy carrying out her orders.

"Found some steamers that were tied up in New Orleans," Natalia said, "and they are pulling out of dock right now. They will not be able to move very quickly upstream due to all the water coming downstream, but they can move up the river enough to help out some. The wind is at its lowest velocity the further south one goes, so that will be the best place to start with this rescue effort. Also the water levels are the highest in this region as the water is all flowing toward the Gulf of Mexico, so these people are at the highest immediate risk. This should be able to rescue thousands of people pretty quickly as these ships can hold a lot of folks."

"Yes, at least along the Mississippi," Kristen answered; "but then that is where the most severe flooding is taking place, as it does drain most of the central United States".

"Director," Susan interjected, "we've located some gambling steamers up in Missouri and Iowa, and they have agreed to help; so have some barges that have been moored all up and down the river. We should have about a hundred ships in motion here pretty quickly," she added.

"That will be great," the director said, "but now we need to get the word to the people along the river who are in danger of being flooding out. We also need to only take those people that are in serious danger as we can only hold several hundred thousand on these ships."

"The hard part will be to get the word out," Elawa reported, "as communication is down almost everywhere. And the other problem is that most of these people probably don't realize the extent of the danger they're in due to rising waters."

"Well, we should try radio, shortwave, cell phones, satellite television, the Internet, the worldwide satellite web, and so forth." Kristen added, "Also word of mouth as the boats start along the river might get some people to safety."

"All the usual media avenues are being used now," Paul replied. "Also the first ships are beginning to pick up passengers, and the helicopters are within thirty minutes now of the river. They should be picking up people shortly after that. Besides that, the helicopters will be able to communicate to the people directly as they fly over the flooded areas."

"That's good news. Did you copy all that, Debra?"

"Yes, I did, and that sounds great. How are the boats to the more remote areas coming along?"

"Right on schedule, from the reports that we're getting," Kristen said. "They should be in the window of opportunity in about thirteen minutes. They'll drop the boats, pick up as many people as they can hold, and get out of there. They're about ten minutes away now."

"That's great," Debra said. "Maybe they have auto cam on and we can watch the rescue effort."

"We're in communication with them, and yes they do have onboard cameras. They're switching them on now. Transfer to main screen please."

"Here's the picture now," stated Natalia. As they watched the picture on the screen, they saw the widespread water sweeping throughout the valleys below. The only areas above water were the peaks of the hills where thousands of people were huddled under trees and ledges trying to get some shelter from the storm.

"As they heard the helicopters approach, the people began waving and shouting at the oncoming aircraft. Then they moved toward the incoming aircraft as the helicopters hovered just above the ground right at the water's edge, lowering their harnesses and releasing their cargo of boats. Then the harnesses were used to evacuate the elderly, sick or injured, and children. In just seconds the helicopters were filled to capacity with soaked inhabitants. Off they went with their loads racing the oncoming squall line. Meanwhile the remaining people quickly boarded the boats as the water level continued its pursuit of the crest of the land.

"We're getting a report from the helicopters, Director," came Debra's voice. "Every chopper picked up the maximum load, and several thousand were rescued. Elderly, injured, children, and pregnant women were taken out first. Water decontamination pills were left behind, along with any other survival supplies that were on board. They all got out of the area before the rain and wind picked up again. From the ground we're getting reports that the boats got to the citizens who needed them most; they're filled and floating out into the newly created lakes, which I'm sure seem to be oceans to those on the little boats. All in all, a very successful mission."

"Good to hear it. We need some more good news in with these storms," Kristen said. "Any more word from the riverboats?"

Debra answered, "Yes, they are picking up people up and down the banks of the Mississippi. Most are being rescued by the lifeboats from the ships, while a few others are now arriving by the first of the choppers in from Texas. We'll be able to rescue thousands more with these riverboats. That was really a great idea." Debra then switched the scan cam to the riverboats just north of New Orleans taking on passengers from lifeboats.

"We're getting the picture in here now; this is great to see. How many loads are we able to move per hour?" Kristen asked.

"They're moving several loads per hour now," Debra said, "and we expect the rate to increase as the people understand what is going on and cooperate on loading the choppers and getting off in a hurry at the other end. We hope that we'll be able to get to up to a thousand an hour real soon. Then the only concern will be how many people the ships will be able to hold. The plan now is for the ships to remain near where they're picking up the people, so that when the floodwaters recede they will be able to unload near where the populace lives. Most of the boats and ships had limited water and food supplies on board when they started to pick up people, so if we can unload within twenty-four to thirty-six hours, then we won't have too much trouble feeding everyone and getting enough water to all the passengers. If however it goes much beyond that, we'll have to figure out how to get supplies to them as soon as we can."

"That will probably work as long as we can get enough boats on the river to hold all of the population that needs to escape the rising waters," Kristen responded, "and as far as the water and food situation, I suppose we will just have to see what we can do when the need arises."

"It all depends on the rain now," Debra said. "If it quits soon we'll be able to ride this one out; on the other hand, if it continues to rain, well—I just don't know. We've surpassed all previous records of rainfall now at every single recording station in the south except for Florida south of Orlando and Brownsville, Texas. We have computer projections that show the rain beginning to taper off in the next twenty-four hours, and if that holds, I think we'll be all right"

"I hear you. Let's hope the computer is right on this one," Kristen said, petting her calico friend.

"We need to go now," Debra responded, "but we'll be back in touch soon; we need to contact all units of the Squad and get their latest report. We'll call you back in an hour."

"Director, Helene is on the screen with an update on the Asian situation," Natalia said. "Switching her to big screen now."

"Thanks, Helene do you copy us?"

"Yes, we hear you just fine. We've got the latest update for you on conditions in the subcontinent. First, the storm is officially completely over, and the skies have cleared everywhere across the region. The water is beginning to recede throughout the area also. Our main problem is the disposal of dead bodies. We suffered thousands dead, and thousands of animals were victims also. Cattle, pigs, dogs, cats, horses, and forest animals were all casualties of this calamity. The forests also took a real beating, and it will be years before they recover.

"The bodies are the real problem though," Helene continued. "We are in a race with disease spreading from the rotting corpses. Bangladesh was hard hit along the Ganges, and in Dhaka the tornadoes killed thousands. We're having a hard time getting to some of the victims inside the collapsed buildings. The other problem is that many of the fatalities are just barely covered by silt in the river bottoms and along the banks—not deeply enough to stop disease from spreading.

"We have shifted everyone who is available to the task of finding and taking care of casualties. About the only ones not involved with this at the moment are Squad members working with the injured in hospitals and triages. It's going to take several days at least to get a handle on the situation. For now we're trying to identify the victims and notify next of kin or, if we can't find them, at least notify the local government who can act as a clearinghouse for people searching for missing loved ones. Then we are cremating the majority of the bodies, as is the custom in most of this part of the world. It is the best way to stop the spread of disease and keep a lid on our problems. We've got enough caseloads right now without some new problems.

"Otherwise we're starting to get back to some degree of normality, and as the water recedes, people are beginning to return to their homes. Many of them are finding nothing left of what used to be their homes, and in some cases there is a new river or lake where their homes were before the storm. And in the delta regions there are entire new areas of land formed by the silt washed down the rivers from all the rain. There are even some new islands just off the coast built by the swirling tides and eddies from the tidal surges. Nature destroys and builds at the same time; the cycle of life and death goes on and on." Helene summed up: "Well, I guess that's pretty much it from here for now."

"Sounds like you all have it under control, as much as possible," Kristen answered.

"As much as one can after a week of being pounded by Mother Nature; but I'd say we're pretty much in control again—as much as we're ever in control of the weather. I need to go and will be back in touch in a few hours. We've all gotten a little rest here now and soon will start regular sleep periods for the Squad members. That's it for now. I'll be back in touch in a few hours."

"Sounds good," the director replied and then moved to the world monitoring computer. "How is the rest of the world doing?" she asked Susan, who was now monitoring the other earth environmental problems.

"Here's the update. First the fires throughout the world are all out except for that one still burning in the northwest United States and southern Canada in Idaho and Alberta. It crossed over the border about twenty-four hours ago, driven by strong southwesterly winds caused in part by the hurricanes' extreme low pressure to the southeast. There is a strong high-pressure system in place; very dry air is associated with it and the humidity is below ten percent, which of course is helping in the fanning of the fires. At the moment the forecast for the region is for more of the same until the hurricanes break down and depart. Too bad we can't get some of that moisture from the South up into the drought areas of the Northwest. This has been one of the top ten years for fires in North America, and the year isn't even over yet.

"The floods in Colombia are tapering off for now, and no new dams have broken in the last forty-eight hours. However, the projections are for the rain to increase again before it stops altogether. Cleanup from the earthquakes in the Philippines and central China are continuing right on schedule. There have been only minor aftershocks in both zones for several days now."

Susan turned to another screen. "Wait—we are showing a new fire … this one in South Africa. They had a fire earlier, but were able to contain it fairly soon after it started. We'll get a report on this new fire in a moment. Meanwhile the tidal wave damage in Thailand and Vietnam is cleaning up pretty good; not much loss of life, either human or animal, as we were able to warn them early enough for them to flee the tidal wave. That's about all that's happening at the moment. It's actually downright calm compared to the last two weeks. We'll keep monitoring that fire in Africa. So far the local officials say they have the fire under control and are not requesting any assistance at this time."

"That sounds like a moderately peaceful world compared to what we've been going through the last few weeks," Kristen remarked. "Maybe we'll be lucky and have a tranquil time of it for a while."

"That would be great," Susan answered.

"Any other news from any where that we should be aware of at this time?" the director asked.

"No, that is pretty much everything that's going on around the world," Natalia said, "but it's not uncommon for things to settle down and be calm for a while following the amount of activity that we've experienced over the last two months. It fact, checking back over the last hundred years, that is the typical trend."

"It makes sense when you think about it," Kristen said. "A lot of energy has been expended between these two storms and the one off the coast of South America. Let's hope it lasts for a little while at least."

Right on schedule Debra, one hour later, was signaling

"Debra is on the screen for you, Director," Natalia stated.

"Great. Debra, how's it going?"

"I just wanted to update you on the situation here. First, the riverboat evacuations are working exceptionally well; we've gotten tens of thousands of people already on the boats. The boats closer to dry areas are unloading and returning for more, while the other boats not near any dry land are going to simply have to ride the storm out and wait for the water to recede. We are going to have to work out food and water for the people on these boats who are riding out the storm. Most of them had little to no supplies on board when we asked for assistance.

"Next, we've had three more tornadoes, all near the coast in North Carolina. Wilmington was the only city hit by the twisters. Two of the storms hit the downtown area, while the third one went through the outskirts of the city. So far no word on any casualties or damage reports. The storms have since dissipated and are no longer a threat of any kind, and we should have some information from Squad members on the way there shortly.

"Now some good news," Debra went on: "the wind is continuing to subside all across the region, and we're seeing the first slight signs of the rain letting up. It's still raining hard everywhere, but not quite as hard as it was in some sections. The computer projections are that the precipitation will recede steadily over the next eighteen hours, with only scattered rain after that for another twenty-four hours. So we may be out of the woods in less than two days with this rain. If that happens, the water levels in the rivers will continue to rise for another few days and then begin to go down. So far that's what the computers are predicting and what's happening. We're still going to have a hell of a cleanup to contend with when this is over, and it will take up to eight days before all the rivers and streams are back to their preflood levels. But at least there is light at the end of the tunnel".

"I think you're right," Kristen answered, "just another couple of days and this one will be in the books. And probably in every record book there is, for that matter. Well, keep us informed and we'll be back in touch soon."

"Roger."

"Sorry to break in like this, but we just have had another earthquake in Colombia, between 5.8 and 6.2 on the Richter scale. We are picking up reports of some minor damage along the coast in the Caribbean and rumors of more extensive damage in the interior. We don't have any word yet on the effect on the dams and levees in the flooded region of southern Colombia. We have many Squad members there, but none have reported in as yet. The communication has been limited at best due to the extent of the rain and infrastructure damage to radio stations and satellite transmitting terminals and the interference due to the storms themselves. It has been real slow going there to get back any information—Wait ... we are beginning to pick up something from the interior. I'm switching it to speakers now ..."

"Does anyone copy, repeat, does anyone copy?"

"We read you loud and clear, over."

"Great, who have we reached?"

"This is Squad headquarters in Colorado, over."

"Roger, we copy. Here's what's happening. The situation is critical. We were just getting things somewhat under control when the quake hit. We had three Squad members killed when a mud wall gave way and buried them under twenty-some meters of mud. It really is a shame as we had no losses before that and the situation was improving. Also three more dams gave way that we had shored up to a point where they would have made it barring the quake."

"We're sorry to hear about the loss of Squad members," Kristen answered, "and will send our condolences to their families right away. What happened from the dams breaking?" It had been a costly two weeks for the Squad and was getting worse.

"We just don't know yet. There are reports of several towns being swept away, but we have no hard data or eyewitness reports. We are trying to reach any Squad personnel in that part of Colombia, but so far have been unsuccessful."

"Let us know the moment you hear something. If you could transmit those names now, we will contact their families."

There was a moment of silence as the names of three of their own came over the screen: two men, one woman.

Natalia gasped as she read the name of the woman. "Sarah Fernandez! Oh no! Sorry, sorry … it's just … She was my roommate at the academy—" Markus got up from his post and went over to her and hugged her. She started to weep silently.

"Natalia, I'm sorry about your friend," Kristen said and gave her a comforting hug, "and I'm sorry for her family; but she knew the risk when she joined the Squad, as we all do we sign up. I know that doesn't make it any easier, it never does."

"Thank you, Director." Natalia collected herself and then went back to her station and back to her job. "Another report coming in from Colombia," she reported, her voice firm and steady, hiding her sorrow.

"Put it on the big screen, please."

"On now."

"Does anyone copy? Over."

"Loud and clear. Who is this, please?"

"This is Paulo. Here's the situation: we have more reports coming in from the outlying areas of major damage to buildings just south of the epicenter. It appears that the quake has had two aftershocks and those have collapsed the already weakened structures. The second aftershock was the most destructive, as it was actually stronger than the first quake itself. We are getting data in now from the World Wide Quake center in Golden, Colorado. Here's the report: the first quake was a 5.8, the first aftershock was a 5.3, and then the second aftershock was a 6.2. They all occurred within a few minutes of each other, which explains the confusion on the first report we received.

We've had several weaker aftershocks since then, but they have all registered under 3.2. We are receiving warnings from the Quake Center that the probability for additional aftershocks is high because the stress pressures have not been relieved yet on the fault line that is responsible for this series of quakes."

"So have you issued warnings to all personnel and civilians in the region," Kristen asked, "to stay clear of damaged buildings and away from dams or levees that might be in danger in of giving way?" She already knew the answer but had to ask anyway.

"Yes, and as a matter of fact, it's already happened," Paulo replied.

"Any other reports from the outlying areas?"

"None at the—Wait, here's one now," Paulo responded. "Another dam is showing signs of giving way. The record rains, in combination with the quakes, have really put the pressure on all those old dams. It's actually amazing that more haven't collapsed, considering the circumstances that we've been under the last week. We're evacuating the downstream zone right now, and I'm putting the word out that anyone living below any dam or levee anywhere should seek higher ground immediately. I'm not sure what good those broadcasts are doing when you consider the situation out there. Power is out almost everywhere, most communication links are severed, and even if anyone did receive the warning, it would be hard for them to go anywhere because the roads are mostly wiped out or underwater. The situation is bleak at best for the people of Colombia."

Paulo continued, "On the satellite view you can see that this is a massive storm with the center just north of the Galapagos Islands. The rain in Ecuador has also been heavy, as well as in northern Peru and northwestern Brazil. The Amazon will most likely be rising dramatically over the next week or so as all this water makes its way down the mountains. One good thing is that a lot of the moisture is being deposited as snow in the higher Andes, which will slow the runoff of the water.

"The part we don't know yet is what impact the snow will have on the western side of the divide, which is where a lot of it is falling since the storm is pumping water in from west to east. The normal east-to-west circulation pattern in this part of the world puts most of the snow on the eastern side of the Andes. We have had avalanches and some soil erosion on the western side, as it is not accustomed to much snow, if any. This is certainly an unusual weather pattern. I suspect it is being triggered by the same forces that have caused the record hurricanes and typhoons of the last two weeks in the Atlantic and Indian oceans," Paulo said.

"You could very well be right," Kristen said. "That's how it seems to us also. We'll know more in a month or two after we're able to decipher all the data that we've collected on these two storms since their inception. If there's anything we can do or any supplies you need, let us know and we'll get right on it, Doctor."

"Thanks. We're in pretty good shape at the moment: a shipment of water sterilizing tablets got through before this last quake and before the rain picked up again, and blood supplies are still adequate for the time being. Food could become a problem in a few days if the floodwaters continue to rise, because some of our reserves are in danger of getting wet. We don't have any way to move them with the roads out, and it's raining too hard to bring in aircraft. But we'll keep in touch, and I'll let you know how it works out. Thanks, and out."

"Okay, we'll keep in touch; out for now," the director replied. "Will you raise Debra, please?" she asked Natalia.

"Yes, it will take just a moment," she replied.

"This is Debra, what can I do for you?"

"Debra, this is Director Verba. How is the boat program going?"

"Very well; in fact, we have stopped on the far western edge of the storms, as the rain there is almost ending. We anticipate that within twenty-four hours all the rain will be light enough that any further flooding of higher areas will not be very likely. Low-lying areas will be in danger for days, but then most of that land was already evacuated."

"Good, then I would like to request that you have the group of aircraft in the western region start heading toward Colombia right away. They may need to have some food supplies lifted to higher ground, and maybe these pilots with slightly heavier craft could get through where others can't right now," she explained. "We checked on the aircraft that they have at their disposal in Colombia, and since it is normally pretty good weather there, they don't have any of the big helicopters that we have for use in the more rugged northern climates."

"Consider them on their way, Director," Debra responded. "Also, we could let some others go by tomorrow, if you need them."

"We'll let you know as soon as we have further data. How's the rest of it going?"

"As well as can be expected. The rain continues to decrease, and that is the best news of all. We were so close to turning the entire southern United States into a giant lake it wasn't even funny. Have you seen the infrared pictures from space?"

"Not yet, what do they show?"

"Well," Debra began, "you know water and dry land appear different under infrared. Look, I'll put them on your screen. See that, almost 20 percent of the entire region south of the Ohio River and east of the Mississippi is underwater. That's the highest known flood level ever recorded in the United States or even found in geological records back to when this was all under the sea anyway."

"Wow, I knew it was big, but this is unbelievable." Kristen gasped in amazement at what she was seeing. She continued, "Hopefully it will clear in time to get a straight photo of the water before it recedes so we have a record to compare any future floods. What do the latest cloud cover photos look like from space, by the way?"

"The skies are starting to clear west to east, but it's hard to know when they will clear over the entire region. With so much available moisture around, there will be afternoon thunderstorms for the next week at least, with daytime heating from the sun, and at worst it will stay cloudy for days more. The weather pattern to the west is not really moving, which is why we can't predict with too much accuracy at the moment when it will clear over the southern United States.

"It almost seems that the weather pattern across much of the world has sort of stopped moving. We have that storm off Colombia, which has been there, what, ten days or more; the hot dry air over Australia has been in place for a long time, even though it is their winter or least near the end of their winter; and if you look about the globe, you find the same stagnant weather patterns all over the world. Something we should look into when this crisis is over, don't you think?" Debra asked.

"I hadn't really thought about it until now as you mention it, but there has been a slowdown of weather systems moving in their normal patterns. Yes, we will look into this for sure. Maybe we can learn what's going on and whether there's any way to predict when these things might happen again. I would like for you to join us here at headquarters, when the situation allows, and help in a review of all that's transpired over the last month or so in the worldwide weather patterns," Director Verba stated. "Your insight could prove to be valuable in trying to figure out what's been going on to create such monster storms with so much destructive power."

"I would love to," Debra answered. "Thanks for the opportunity to contribute. My major at the academy was meteorology with an emphasis on climate changes. I don't know if we are experiencing an increase in the climatic changes due to global warming or if this is an isolated weather phenomenon, but it sure is something that we'd better try to understand either way. If it is a pickup of climatic change, then we've got to be prepared; if it's only a once or twice type of thing, we'd still better be ready for when it strikes again."

"You'd better believe it," Kristen said. "We need to understand what's happened. But anyway, thanks for getting those helicopters in route, and any others that you can free up could be helpful. We'd better keep some of them in this part of the world in case something else happens like a dam breaking or whatever. It might be useful to have some close by, just in case."

"Sounds like a plan," Debra acknowledged; "we'll keep about half here and hopefully be able to send the rest to Colombia as soon as possible. Over and out for now, we'll be in touch soon."

"Okay, talk to you soon," Kristen answered and then asked Natalia, "Have we heard from Helene recently?"

"No, but we can try to raise her if you would like," Natalia responded.

"Please, we need to see how it's going in India."

"Stand by, we're contacting her now."

"This is Helene, how can I assist you?"

"Hi, Helene, this is Kristen. Just checking in—how's it going?"

"We're doing pretty well. We've managed to retrieve a lot of the casualties, and we're getting that situation somewhat under control. We still have some problems with the partially buried bodies— locating them and getting them to a disposal site. But we're doing the best we can under these conditions; with any luck and a lot of hard work we'll be successful. The lack of passable roads is a major obstacle, and the vast areas that were flooded is the other significant difficulty we have to contend with to find the remaining bodies. Along some of the rivers, three meters or more of silt and debris covers the surrounding land. We are having trouble getting through the silt without submerging our people or vehicles. In some cases we can see parts of some of the victims sticking out of the mud, but we can't get to them yet. Also there are the animal victims that are partially submersed that we will need to dispose before more disease spreads. It is pretty gruesome work.

"We have a lot of help from the local governments," Helene went on, "but it is too large a job for their already strapped resources. There is so much widespread damage to government buildings, vehicles, communication facilities, and so forth that it is hard for them to mount a major search and rescue effort until they get their own people located and start repairing their devastated equipment. It will be slow going for a while, but we should start to see some progress soon as new equipment and parts to repair the salvageable vehicles arrives. The lack of parts is a big problem; many of the vehicles are old, and there just aren't readily available parts. We are substituting wherever we can and we've got the parts and procurement people on it around the clock. They've been just great getting parts to us here. We had a shipment airdropped in a little over an hour ago. They have sent people to look at the equipment and ascertain what can be done to fix it. Send our sincere appreciation to that part of the Squad."

"I'm sure they will say they're just doing their job, but consider it passed on," the director replied.

"Thanks, they deserve a little recognition every once in a while. They usually work behind the scenes, and we sort of take them for granted, if you know what I mean," Helene said. "Anyway, we will be putting most of our effort into the collection of victims until we are finished and, of course the treatment of people who need our help. Next we are providing food and water to wherever it's required and then we are starting the rebuilding process."

"That sounds like a plan. We'll be back in touch with you in a couple of hours. Out for now," Kristen said.

"All right, I'll talk to you soon."

CHAPTER FIFTEEN

There was a lull in the action, and Director Verba went back to her office, followed by Silverton. When they got to the office, the calico went outside through her cat door. Kristen had been worrying about her mother and took this opportunity to call her. As she had guessed, her mother was taking her uncle's death very badly. They talked for almost half an hour, mostly reliving all the great times they had all had with her uncle. Those had been great times, and they both fondly remembered their lost relative. Her mother thanked her for calling; it really had helped cheer her up quite a bit. They made some plans for Kristen to come visit as soon as she could spare a little time from the Squad. With a page coming for her to return to the control room, she said good-bye, promising to appear at her mother's door as soon as events allowed. Silverton seemed to know that the director was through with her phone call and appeared just as she hung up. The two of them headed back to the control room.

"Director, some good news," Natalia greeted her, as they entered the room, Silverton climbing to her usual perch. "The storms in North America are finally letting up for sure. Both the wind and precipitation are decreasing significantly now throughout the entire continent. And on top of that the rain in Colombia has started to diminish also. There is light at the end of the tunnel after all," she finished.

"That's great!" Kristen said, truly enthusiastic from the good news, plus the uplifted feeling she had from talking to her mother.

Things were embarking on what would hopefully be a better time for the Earth. Continuing, she asked, "How about the earthquakes, fires, and those other problems that we've been having?"

Susan, monitoring the rest of the world, began her report. "First, the earthquake regions, except for Colombia, haven't had a significant aftershock for days now including the China, Japanese, and Philippine quake centers. Relief efforts are well under way in those sites except for the Colombian area because of the rain that has fallen since almost the beginning of the first earthquake. Now with an improving situation, we hope to get rescue personnel into place real soon. The fires are mostly in check with the exception of the North American fire between the U.S. and Canada. That particular one is burning out of control still with no immediate relief in sight, and the wind pushing it further into Alberta and now also into Montana. We hope to make progress within twenty-fours, though, with major fire lines and back burns that we'll be attempting tomorrow. If that works, it will still be a couple of days at least before it is completely out. If it doesn't—well, we'll have to think of something else." Susan sighed. "One major problem is that it's so dry right now in that part of the world. Too bad we can't get some of that water from the southeast to the northwest."

"That would be great if we could," the director responded. "Anyway, do we need to put more Squad members onto those fire lines and the back burn attempt?"

"We have quite a few Squad members in place, and the line is pretty much ready. The problem is the wind; it's picking up embers and blowing them across the fire lines that we've tried so far. It has started one fire almost a kilometer past our line," Susan answered. "The plan is to have a line at least two to three kilometers wide and back burn from there toward the oncoming fire. We're also hitting it pretty hard from the air; however, we did send many planes to the southeast and to the Colombian situation. We should have more back now as the situations subside at those sites. There are also all the other fires in this region that, although they're under control, aren't out completely yet, and we're still pounding them from the air as well. When those fires are declared officially out, all those planes will be transferred to the big one. So more Squad members aren't what we need; what we need is for the wind to quit blowing so hard."

"I see what you mean. Let's hope that works."

"It should with any luck. We've been pouring fire retardant on the fire, giant machinery has been cutting a huge swath of barren land between the fire and virgin forest, and the back burning has already started. The wind shifted for a moment, and a window opened for the back burn to begin. In fact, I'm seeing the report come over the computer right now … Let's see, they're starting about a hundred fires along the line…the first ones are beginning to start back toward the oncoming inferno," Susan said as the information poured into the control room.

"Keep us posted, and I'll check back with you in a while. Thanks, Susan. You're sure getting your feet wet here in a hurry."

"Yeah, it's not how I expected to spend my first couple of weeks at headquarters, but it sure has been an education, and I've seen some things that we might be able to improve on, from this firsthand experience in a major world disaster. I just hope I've been useful," she stated.

"Of course you have, we all have, and in fact, everyone in the Squad has been doing a bangup job through these last few weeks. I'm proud to know all of you and to have this opportunity to work with such a great group of people," she stated, moved by how her fellow Squad members had performed under such extreme stress

"Director," Natalia interrupted, "Helene is on the line for you; I'm putting her on the big screen now."

Focusing her attention to events outside the control room, Kristen turned to the screen and said, "Thanks, Natalia." And then to Helene, "Helene, how's it going?"

"Pretty good. Our rescue efforts are in full swing everywhere in the region now. We've been able to get planes, helicopters, boats, and ground transportation through to the stricken zones. The road repair effort has been going slowly, but we are starting to get the main traffic routes back in operation. We've treated the majority of the wounded by now that have come or been brought in for help. We are still finding people who require some help, but even that number is dwindling. Food and water supplies are increasing steadily now as transport after transport arrives with procurements.

"The recovery of victims though is still going slowly," Helene went on. "We are trying to identify as many people as possible, but that is proving to be very difficult in quite a few cases as decomposition has already commenced. In those cases we have no real choice except to take a DNA sample for later identification and quickly cremate the bodies. This is by far the hardest part of our task. But all in all things are improving, albeit slowly. By the way, who's still in the World Cup?" she asked, trying to put some degree of normalcy back in their lives.

"Well, I think our entire group has made it to the quarterfinals plus one other team, I think it's … just a minute, here's the chart on the quarterfinal games. Australia plays Canada, Botswana plays Brazil, Sweden versus India, and the United States plays Estonia. Then the winner of the Australia–Canada game plays the winner of the U.S.–Estonia game, while the winners of the other two games play each other. And, of course, the next two winners play for the World Cup. Well that's where we stand as of now. The first games are tomorrow in Russia the host country for the finals this time."

"You know, Alex was born there," Markus said looking up someone in the Squad who was from Estonia. He was thinking that it would be interesting to have someone from all the quarterfinal teams represented in Squad headquarters. "Maybe we can confirm he is from Estonia?"

"Sure, we need to check in on the Florida situation anyway," the director said. She told Helene to keep in touch and to call headquarters right away if any needs arose. Then before reaching Alex, they called Debra as the situation was still more critical in the rest of the southeastern United States than in Florida proper, which had been out of the brunt of the storm for several days now.

"Debra, how's it going?" Kristen inquired.

"Well, the worst of the rain and wind is over, although we still have gusts to just over a hundred kilometers an hour in some places along the Atlantic coast. Daria never did quite merge back with Colin and Eric and moved more inland to the northwest over Texas, Oklahoma, Kansas, even west into Colorado and is raining itself out. It has weakened a great deal and spread out over much of the central U.S. You're probably getting some effects there at headquarters. Rain is falling right now in fact from Colorado to the East Coast and from the Gulf of Mexico to the Great Lakes. In some ways it's helping that it is covering a wider area; that has the effect of spreading the moisture over a much less concentrated region.

"But we still have a huge, and I mean *huge* area underwater," Debra went on. "The good news is that the water is starting to recede slightly. We are anticipating that some people will be able to start returning to their homes within a few days, with most back within a week, maybe a bit longer for some scattered regions closest to rivers, the coast, and areas that were below washed-out levees and earthen dams. We're still picking up a few scattered tornadoes, mostly along the East Coast, and so far we've been really lucky; no reports of any further damage. The twisters have been near the water and have all moved out over the ocean before causing any destruction of property or loss of lives."

"There does seem to be a light at the end of the tunnel," Kristen said, with Silverton now on her shoulder. "So all of those tornadoes moved offshore?"

"So far that's been the case, and we hope it continues. Wait ... we are picking up another report from the coast of Georgia ... Here it is, and we're switching to audio—"

"Debra, this is Squad member Leslie Stewart reporting from near Savannah. We are reporting several tornadoes right now in the center of the city. Other Squad members are on the scene already; I'm picking up broken transmission, I'm putting it on audio: … We … we're—there they are! We've got a band of tornadoes moving into the city of Sav—annah … We—" The airwaves went silent.

"Did we lose them or what?" Kristen asked.

"I don't know … we had them, and then nothing. We're still trying to reestablish contact, but nothing so far. All we can do is keep trying."

"Well, let us know as soon as you reestablish contact," the director instructed.

"Will do, out for now."

"What is the situation in Colombia?" Kristen asked Paul who was monitoring that situation.

"The good news is that the rain continues to abate," Paul reported, "although it will be some time before the water recedes enough to expedite rescue efforts in the region. And then the other good news is that there have been no aftershocks for nearly three days. We're hoping that the worst is over and that the rebuilding effort can start soon."

"That is good news. Now if we can only get the North American situation under control, we'll have the entire world back to some degree of normalcy," the director stated as a matter of fact, with her calico cat on her shoulder.

"Director, look at this," Natalia said from the far side of the control room. "The fire in Alberta, Idaho, and Montana is coming under control, and with any luck will be by noon local time," she continued while switching the information to the big screen.

"That is really good news. How are all the other fires doing?" Kristen asked.

"The Alberta, Idaho, and Montana fire was the only fire in the world that was still out of control at the moment," she answered. "All the rest are either out or just being mopped up."

"That is great news. Maybe we can have a couple of days here when nothing is going on somewhere in the world," the director responded back to the Brazilian.

"Cross fingers and touch wood," she replied with the old superstition, continuing, "let's hope so."

"Director, we're back with Savannah, switching to large screen now," Susan announced.

"This is Squad member Mario Chuttzi, does anyone copy?"

"We hear you loud and clear. This is Leslie, go ahead."

"We were hit by the tornadoes and lost communication for a while. We don't know the extent of the damage at this moment, but we have teams headed into the area. The tornadoes have moved off the coast and are no longer a threat to anyone. Wait, I'm getting the first reports from the region hit … There are quite a few buildings down … There are bodies lying throughout the zone … some scattered fires … many trees toppled … cars strewn about … Loss of life is definite, we need to set up triages right away to treat the wounded. Can you send medical teams into the area as soon as possible?"

"They're in route, along with blood supplies, food, and water. Is there anything else we can help with you with?" Leslie asked.

"That should do it for now. We'll let you know if anything else is needed," Mario responded from the field.

"Did you copy that, headquarters?" Leslie inquired.

"Yes we did copy; keep us informed if there's anything we can do for you. Talk to you soon, out."

"Will do Director, out for now."

"Director, check this out," Natalia interjected.

"What's happening?" Kristen asked.

"Thought you would like to see some really good news. The fire in northwestern North America is finally under control with only scattered areas still burning, and those regions are expected to be out within twenty-four hours. The wind shifted again, and the back burning headed right into the main fire, which burned itself out way ahead of predictions. So for the first time in seven weeks we have no fires out of control on the planet."

"That is good news. It seems that maybe we have turned the corner, at least for the time being. How's the situation in Colombia, Markus?"

"The rain has definitely abated," he answered, "so for the first time in days rescue teams will reach some of the more remote locations that were hard hit by the earthquake and then the floods. That's the good news; the bad news is that we are finding many dead, many more missing. Thousands are injured and we are setting up triages throughout Colombia and the wounded are being dealt with posthaste. We have already administered aid to some thirty-eight thousand people, and there are some eighty to ninety thousand still waiting to see a medical attendant. Most of the injuries are minor, so we are treating them and releasing them as fast as we can move them through the system. The ones who are injured more severely are being airlifted to hospitals in the north where there is more bed space. The beds we have in the south are being reserved for the those who can't be moved without further possible impairment. We're in pretty good shape at the present time for medical and blood supplies. We could use some more drinkable water in some of the more remote regions in the southern part of the country, but otherwise we're looking good for now."

"Where is the closest drinkable water for the southern region of Colombia?" the director asked.

"We have some in route now from Brazil and Argentina," he answered, "and more tablets for purifying water coming from Chile. It all should be there within a few more hours, as it left a while ago."

"That's good to hear. If anything changes, let me know right away," Kristen responded.

"You've got it, Director," Markus replied.

"Director, it's Debra on the big screen," Natalia interjected, and then continued, "She has the latest reports on the situation in the southeastern United States."

"Thanks. Debra, what's up?"

"Director, here's the most current situation as we know it: first those tornadoes that we were monitoring have dissipated. However, new ones have developed over Chesapeake Bay and are moving toward Virginia Beach and the navel station located there. Right now they are headed directly at the navel facility, and they seem to be twelve to fifteen in number at this time. We don't have any Squad members on the scene to tell us exactly what's going on, but triple Doppler radar indicates that the twisters are moving in a band toward the main shipyard and that they should hit within about thirteen to eighteen minutes.

"We, of course, have sent many warnings to everyone in the region. We assume that most people have taken shelter from the storm, if they have heard our broadcast or other warnings going out over all the different official channels. We haven't received any acknowledgements from the base, but our sensors picked up that they did copy the transmission and are most likely trying to reach shelter before the storms hit."

"That's probably what's going on," the director said. "How many tornadoes did you say there were?"

"At first we thought there were somewhere around twelve to fifteen, but now we are showing over twenty-five, and it could be as high as thirty-two. They keep forming and disappearing rather fast, so it's hard to pinpoint the exact number," Debra said.

"I see. Well, it looks like they're about to hit, can we switch to one of remote cameras in the region?" Kristen asked Natalia.

"They're coming on the big screen ... now," she responded. "These are at the naval yard, and you can just barely make out the tornadoes in the top left-hand portion of the screen. They are moving on a path at the moment that will make a direct hit on the yard in about eight minutes. The cameras are pretty still with the calm before the storm."

In less than a minute the cameras began to sway and in two more minutes were shaking violently in the oncoming twisters. The room went completely silent except for the soft purring of Silverton, asleep on her favorite computer. They all watched as the next five minutes seemed to grind to a halt, moving in an eerie slow motion, than in a burst of incredible power; the cameras showed buildings bursting into splinters, accented by numerous lightning flashes and then darkness as the cameras were destroyed by the fury of the storms.

"Switching to another scan cam," Markus stated and the picture returned to the screen on the wall. These cameras were near the storms, but not in the line of tornadoes. They did show the fury of the twisters, as the wind was blowing extremely hard.

"Where are the storms now?" Director Verba asked.

"They are approaching the coastline and should be over water in a few more minutes," Natalia answered. "That should reduce the risk to any more land areas unless they veer back to the northwest."

"Let's hope that's the case. Debra, are you still there?" Kristen inquired.

"Yes, I am still here."

"Debra, can you hear me okay?"

"Yes, loud and clear."

"Do you have any word from anyone in the tornado's path?"

"Well, it has only been about thirty minutes since the tornadoes hit, so they probably have been assessing their situation, but we have had some broken transmissions. It seems that their communication systems were damaged, but they have backup procedures that should be able to bypass any nonfunctional equipment and get the whole thing back on line pretty quick here. In fact, they're coming back on now. I'll patch their transmission through to headquarters. Do you copy them?"

"Yes, can they hear us?"

"We read you," a voice said over the airwaves. "I'm Chief Petty Officer O'Brian with the Naval Shipping Yard here; and I'm sorry to report that we've had some casualties. We have at least three dead, twenty-eight injured, and over a hundred unaccounted for at the moment. We're hoping that most of them are just in hiding and will show when the all-clear is given. There was very little warning that these storms were coming, and most everyone just had to seek shelter in the closest place they could find. I know that some people were already in shelters from the hurricane, while others were on limited duty, monitoring crucial areas of the yard. Those individuals are the ones missing. A couple of the missing reported that they hid in the lower compartment of a battleship. At least they were pretty much safe there. So we're about to commence a building-to-ship search as soon as we can safely go out. Wait—I am being told that we have more personnel checking in. They're starting to trickle in as the weather improves. But as you know, the hurricane is still hitting here with some force, although not anywhere near the level it was even a few hours ago."

"Yes, we show your winds at still near hurricane strength, but definitely abating," Debra interjected.

"One moment please, I have people updating here. Here's the report I just got. We have found another seven dead and six badly injured in a collapsed building. We're sending medical staff double-time. Another account of a small plane hangar being destroyed and four more bodies found there … Here's a good story: three found alive inside a closet that was the only part of a building left standing. They were in the direct path of the twisters, and it's an absolute miracle that they're still alive. Here's a picture of that building we're getting from a scan cam that managed to escape the worst of the storms," Chief O'Brian said as the picture came up on the screen in headquarters.

"*Wow!* They were lucky! Check out that building," Markus said when he saw the pictures.

"That is completely amazing that they are still alive," the Director said as she looked at the building, which was totally destroyed, save for one room: the closet. Every other aspect of the building was gone; the walls, the roof, all the other rooms that were there, were all gone. The door was open and the three inside came out, finally free of the fear that had totally consumed them for the last hour. The tornadoes had been gone for an hour, but they were afraid that if they tried to open the closet door, the remaining building might collapse, so they stayed in the closet until a passing rescue team discovered them.

"They appear to be all right. See if we can be of any assistance," Kristen said, then went on, "Debra, are you still there? Can you hear me?"

"Yes, Director, loud and clear. What can I do for you?"

"Can you update us?" Kristen asked.

"Well, except for the tornadoes that just hit Virginia, we're pretty much under control now. The rain is still falling, but the rate has decreased quite dramatically over the last twelve hours. I think, or I at least I'm hoping, that the worst is over. And computer projections are confirming what we're observing. We could give the all-clear as early as tomorrow night. Of course there is still a chance that the storms will go out over the ocean and return inland. But all the computer predictions at this time indicate that the storms are going to move out to sea and fade away."

"That's some of the best news that we've heard for a long, long time. Do you have the current odds that it will turn back toward land?" the director asked.

"At this moment, the computer is showing only about—let's see … 15 to 23 percent," Debra said, at last finding the numbers on the screen.

"We'd better put at least a level-one watch out for the upper East Coast of the United States," Kristen said, "until the storms pass completely out of here. We don't want to take any chances on any more lives lost now, we've lost enough, more than enough."

"It's already done, Director," Natalia said; "we've had a level-one warning and watch out for the entire eastern seaboard of North America from Virginia to Newfoundland and Greenland for a while as a normal precaution. We really expect the storms to move toward Europe, but to have lost any punch way before they reach the European continent. Anything else?"

"Not at this time, but keep us informed if anything changes, O.K.?" she said.

"Director, we have a call coming in from Helene," Markus stated. "I'll switch it to main screen now."

"Thanks. Helene, what is the situation?" the director asked.

"Just thought you would like to know that most of the rivers are starting to subside, with the exception of the Ganges, which is at least not rising any more. It was still rising until just a few hours ago as it drained all the other rivers and streams in this huge watershed, which took the brunt of the storm. Now for the first time since the storm approached land a few days ago, although it seems like weeks ago, no rivers are rising. This has been the longest week I can ever remember," an exhausted Helene stated.

"We all know what you mean," Kristen responded. "In fact, the last two to three weeks have seemed like an eternity. We're all very relieved that it finally appears to be ending. The last report from North America put the eye of the storm leaving the East Coast of the United States and moving out to sea. There is still a very small chance that it could regenerate and come back onshore, but the computer models don't indicate much probability that at the moment."

"Well, that does sound good," Helene said; "at least we're close to the end of this … this—I don't really know what to call it."

"Yeah, it has been one heck of a week. Hopefully there won't be anything like this again for a long, long time," Kristen responded.

Silverton, who had been asleep on her favorite computer, woke up, stretched, arched her back, and walked over to the director. She waited to be picked up and immediately went to Kristen's shoulder and started purring just as another message began to come in to headquarters.

"Director, Debra's on the screen for you," Susan announced. "Switching now." Debra's face came up on the main viewer.

"What's up, Debra?" the director asked.

"Here's the current update: we've got every river, lake, stream, you name it, in the entire southeastern United States over its banks," Debra reported. "We now have confirmed thirty-five hundred dead and over thirty thousand missing. Thirteen thousand are missing from that one town in Arkansas and feared dead at this time. We have nearly one million injured and over thirty million estimated homeless right now. All these numbers are expected to rise over the next few days as the water recedes and we start to get into all the remote areas that have been cut off from the outside world. The good news is that the water is starting to go down in some places and that the wind and rain have abated quite a bit over the last several hours. We expect that the worst is finally over and that we will be out of the woods soon."

"Is there anything that we can assist you with at this time?" the director inquired.

"Not right now, although we've sent a preliminary list of post-storm rebuilding supplies that we'll need as soon as we can get into the affected areas. Roads will need to rebuilt first so we can get into regions that were cut off, especially in the hills of Arkansas, Missouri, Tennessee, and all the way through the Carolinas. Parts of Mississippi, Alabama, Georgia, and Louisiana have not been heard from in days. We will get helicopters in there as soon as the weather permits. We'll drop in medical teams and supplies and then work on getting anyone out to hospitals who might need medical assistance. After that we'll be working on ground routes into these areas to get major relief efforts under way.

"Actually," Debra went on, "we've already started with the helicopters in the western portion of the storm area. In fact, we have reports from Arkansas now as the first choppers approach the town that was annihilated by the breaking dam. This is the first time that we've been able to get aircraft near the town, as the weather over the top of the mountain has been shrouded in clouds until right now. They're switching on their cameras, and you should be getting pictures now."

"We've got them." What a sight; everything was gone: trees, buildings—only remnants of foundations were visible. "What a nightmare this must have been," the director stated, barely believing what she was seeing.

"This is chopper five, do you copy?" the pilot asked.

"We hear you, go ahead," Kristen responded.

"We are going to follow the valley down the path of the water until we find something," the pilot continued. The pictures showed the path of destruction as it zigzagged down the winding valley taking out everything in its path. Then the chopper came around a bend, and the valley spread out dramatically, which slowed the power of the once concentrated wall of water. There were buildings, cars, trees, heaps of debris as the wall of water lost its power but left its mark on the mountain valley. Then they saw that there were people on the tops of some of the buildings waving to the chopper as it flew over.

"There's survivors!" the pilot shouted as the chopper set down near the people. "We're seeing at least a dozen—wait, more are appearing ... We'll get back to you as soon as we get these people on board," the pilot said and signed off to get the people loaded.

The chopper cut the blades as several survivors ran toward the helicopters. They had lived by floating down the rushing water on the tops of their buildings. It was a miracle that the buildings remained upright and that they could hang on in the turbulent waters. It quickly became apparent as the crowd grew that more people had survived than could fit on one chopper, and the word went out for more aircraft to come immediately and help with the rescue effort.

"Hello, headquarters, I have been talking with the people here," the pilot interjected, "and they told me that some thirteen thousand people lived in the town when the wall of water went through, and now it is beginning to appear that some two hundred or more have managed to survive the water at this site alone. Maybe there will be more as the search continues."

In just about thirty minutes more helicopters were landing around the survivors and began to take on the lucky ones. The pilot continued relaying the stories he was being told. "The people told me the water came and went so quickly that anyone who could hang on for a fairly short time was able to survive. They said it was like riding a surfboard or snowboard. Others seem to have managed to get to the edge of the water and reach a landfall where they were able to get to higher ground. Still others just swam to safety, an amazing effort of personal strength and determination. It's utterly astonishing that anyone managed to escape the fury of the water, let alone the apparent hundreds who are being rescued. This is good news."

"This is chopper one. We are full and are heading back to base with our wounded on board. We will drop off these passengers and return to the site as fast as we can to get more people out of here. We took the most critical on this first load and those remaining are in pretty good shape for the most part, save for some dehydration. We left all the water we had and will pick up more when we return. That's our report for now, over and out."

"Thanks, chopper one; we copy and concur with your plan. Keep in touch," the director stated.

"Roger that," chopper one answered.

"Debra, did you copy all of that?" Kristen asked.

"You bet we did; that was good news. We didn't expect anyone to make it through what happened there, so any survivors are great news!"

"You bet it is! Can we get any more help into that region?" the director asked.

"We have sent every available helicopter within a thousand-kilometer radius to the area, to help with the rescue and recovery effort. We should have about a hundred choppers in the region within another three hours or so. That should be enough to get everyone out of there and to a hospital if they need it or at least to safety."

"Sounds good."

CHAPTER SIXTEEN

The rescue effort in Arkansas continued for the rest of the day. When all was said and done, there were 853 people who were found alive from the dam breaking and the resulting wall of water that careened down the valley. With nearly thirteen thousand in the valley, there were still twelve thousand who were missing. A few bodies had been found, but most were still missing.

"I'm afraid we will never find all the missing persons, as they're most likely under meters of mud and silt," Kristen said after seeing the number of found bodies.

The situation in Asia was improving every day. All the rivers that had flooded over their banks were now below flood stage, and although there were still millions missing, most were thought to be just separated from their loved ones and not dead. The storm had made almost a hundred million people refugees from their homes, so the possibility that most of the missing were only somewhere else was quite high. Every day another several thousand were accounted for, and the number of bodies being found was dropping off. The Squad members in Asia began to shift the majority of their effort to reconstruction now as the rescue effort diminished. Infrastructure was the number one priority. To get the roads, bridges, and airports rebuilt was first on the list to accommodate rescue and relief efforts. Silverton was in her usual spot—on the main control center, curled up into a furry ball, purring away when the call came from Debra.

"Director, do you copy?" asked Debra.

"We copy. What can we do for you?" the director responded.

"Just wanted to update you with the latest information on the storm. This is very hopeful, but I think it's over! The center of the storm has drifted nearly three hundred kilometers off the East Coast of the United States, and it also seems to be weakening as it moves into cooler waters further north in the Atlantic. We actually have clear skies in Texas, Arkansas, Missouri, and parts of Tennessee. The clear skies are expected to move east as the night goes on, and by morning we could see most of the southeastern United States with clear skies. It's over. Now we just have the rescue of stranded individuals, the cleanup, and the rebuilding ahead of us."

"We made it! It could have been a lot worse without the evacuation effort that we had prior to the storm. All that work paid off, although we still had quite a few casualties. But it would have been much worse if we hadn't put so much effort into getting people out of the danger areas. Think about what would have happened if we hadn't done anything at all and no one had been pre-warned of the impending disaster," Kristen said excitedly.

"Yeah, no kidding, it could have been way more tragic. Still, we lost several tens of thousands, but it would have been worse if we hadn't forced such a massive evacuation prior to the storm," the Debra stated.

"And the effort of the helicopter pilots, ship captains, and the millions of unsung heroes in these past few days…well, it goes without saying that they were all instrumental in getting millions who might have been lost to safety," Kristen said.

"That is true," Debra said. "We will never know most of them by name, but they will know who they are and what they did."

"Yes they do," Kristen said, "and I hope they're really proud of what they have done over the last couple of weeks; but I'm sure they think it was just their job. Anyway, is there anything that we do help your effort at this time?"

"We've sent a list to Procurement, and they're working on it," Debra replied. "A lot of heavy equipment, food, temporary shelters, water purification tablets, and of course some more medical supplies. We are still treating the injured and are finding more all the time as we move into areas that have been cut off since the height of the storm. Tornadoes caused an extreme amount of damage throughout the southeastern United States. And we can't forget the Bahamas also suffered major destruction."

"Yes, we can't. Anyway, we'll talk to you in a couple of hours, and try to get everyone some rest, okay?"

"Actually we've already started getting some sleep shifts for Squad members," Debra explained, "and by tomorrow everyone will have had at least one sleep shift. By the next day we expect to be back on regular shifts for all personnel, with the exception of some medical teams that are still very busy. However, even those Squad members are expected to get back to regular shifts within a few days."

"That sounds really good. Well, take care, talk to you later."

"Out."

"Natalia, what's the latest on the rest of the world?" Kristen asked.

"The fire in Africa, more specific South Africa, is under control, and the last mopping-up effort is under way now," Natalia reported. "The situation in Colombia is steadily getting better as the floodwaters recede, and there have been no major aftershocks for a few days now. The situation in the remote areas is still critical, as it is very difficult to get any sort of relief effort into those regions. We have helicopters coming in from the southeastern United States, which will be in position within another few hours. It will take several more days before all the injured are taken care of; we may never account for all the missing, and never know the names of many who are missing."

"How's the effort in the Philippines and southeastern Asia going from the tidal wave damage?" Kristen asked.

"That is coming along quite well. Because of our evacuation efforts the loss of life was kept fairly low, but the amount of physical damage was quite severe, especially to the infrastructure and public buildings. Hence the majority of our effort is the rebuilding of roads, bridges, airports, docks, hospitals, schools, and so forth. We hope to be near 50 percent rebuilt by the end of the month, but then the other 50 percent will take longer; it is quicker to repair a road or build a bridge than it is to finish a hospital or airport, for example. We have worked first on the reconstruction of major transportation routes to facilitate the movement of needed goods and supplies into the stricken regions and then ancillary on public buildings.

"The Philippines are recovering from their earthquake in pretty good shape," Natalia went on. "Spirits are high as damage, although very extensive, did not cause a great loss of life. A great deal of that fortunate situation was caused by the fact that a large percentage of the population was in the streets celebrating their country's win in the World Cup. Even though they would go on to lose in a later game and be eliminated from further competition, it was a very important win to the people of the Philippines because of the number of lives it saved because they were not indoors.

"The earthquake in China has been followed by several aftershocks; however, none have been strong enough to cause any major damage, only enough to keep people on a slightly nervous edge. That pretty well sums up all of the world's major situations, excluding the two hurricanes, Director. Any questions?"

"No, not at this time. You covered everything I wanted to know about these other situations, and it seems that the circumstances are improving worldwide. I think that we can start to get back to more normal working hours here," the director instructed, "starting now. Let's have half of you take an eight-hour sleep shift, and when the first half returns, the other half can take the same break. After that we will go to a level-three alert and go back to regular shifts of duty twenty-four hours after that. You all decide among yourselves who wants to go first, and we'll see you back here in eight hours."

"I'll take the first watch on," Markus stated first.

"I'll stay on too," Natalia said.

It was soon decided that Susan, Paul, and Indira would take the first sleep shift, and Markus, Natalia, and Elawa would take the second one. The director stayed for the first four hours of shift one, and left stating that she would be back at the outset of the second shift.

As she left, her cat jumped down off her computer and trailed behind Kristen back to her office. Then the director called her parents in Australia. They were recovering from her mother's brother's death the best they could and insisted that she visit them as soon as possible. She sensed in their voices that they weren't doing as well as they pretended. She agreed that she would visit as soon as she could get away, after all the current disasters were under control. In her mind she knew that would be at least a month or two—or three. Anyway, she thought she could probably make it home before Christmas. She then lay down on the cot in her office and in moments was fast asleep. Silverton crawled up on the cot, curled up behind the Director's bent knees and fell asleep, softly purring.

The first group left to get some sleep, and the others settled into their watch.

"It seems so quiet in here," Natalia said. "With less people in here and the big screen silent, it seems so calm."

"It really does. It has been so full of activity in here for weeks, that this really seems extremely quiet. But don't get me wrong, I like it quiet after all the world has been through the last three weeks," Markus answered. Just then a call came in.

"This is Debra, do you copy?"

"We hear you, what can we do for you?" Markus asked.

"Is the director there?"

"No, she's sleeping right now. Do you need us to wake her?" Natalia inquired.

"No, we have only good news at the moment, so she can sleep for a while longer. But when she does come back, tell her we have found another eighty-five people alive in Arkansas from where that dam broke. They managed to get up a large oak tree near the edge of the water that was slightly up a hill, which slowed the wave just enough for them to escape. The water ascended about five meters up the trunk of the tree, and people were hanging on everywhere, from what they told us. We found the survivors huddled together near the tree. Most of them were in shock, and some had some bruises from falling out of the tree along with a couple of minor scrapes, but otherwise they were alive, and all of them will be just fine. Plus we have located another twenty-three people up and down the valley along the floodplain. To have found anyone alive after the amount of water that must have come down on that town is truly amazing.

"Other miracle stories are trickling in from around the nation. A little girl and her dog were found in a basement of a house that a tornado had completely destroyed. The dog had kept her warm. Her parents were next door helping their neighbors give birth when the twister hit. They didn't have time to get back to their own house and went into their neighbors' basement. They could only hope that their daughter was safe. It has taken them until today to dig through the rubble to get to her. They had to do it all by hand, because all the roads are washed out, keeping all the emergency equipment out. They got to her just in time because the basement had flooded and the water was still rising when the found her. She seems pretty much okay though and is expected to fully recover within a day or so. Anyway, these are some of the stories that are coming in to us here at regional headquarters. Thought you all might like to hear some good news for a change."

"Yes, we appreciate hearing some happy endings after all the sad news that we've heard over the last three weeks," Natalia responded. "There will be many hurting families throughout the world, so any happy news is great." Then she changed the subject and asked, "How's the weather doing?"

"We show it clearing from west to east and south to north," Debra answered. "We expect the entire region to be clear by tomorrow. This is great news too, because if we received any more rain at all—I just don't know what might happen. We already were at record flood levels everywhere in the southeastern United States, so any more rain would have been a disaster beyond anything we could imagine. Now we're showing rivers, streams, and lakes all beginning to recede. Some people will be able to go back to their homes within another couple of days to at least check on their property and move back if conditions permit. Most will be able to check their properties by the end of next week."

"Good news for all those inhabitants," Markus declared.

"Yeah, and by the way, how's the situation in Asia?"

"Well, things there are going pretty well," Natalia answered, "considering what they went through. The skies are completely clear throughout the region, the waters are receding in the flood zones, and we've located the majority of the missing people, which as we suspected were only separated from their loved ones.

"The bad news is that we have found more than two hundred thousand dead from the storm. Mostly due to flooding from broken dams, tornadoes in Bangladesh, and storm surges along the coast. The amount of farmland buried under meters of debris is staggering. The storm destroyed large areas of forests, national parks, and wetlands valuable to waterfowl. It was an absolute disaster for wildlife. We found many rare animals just roaming around outside protected areas. We've had to bring special endangered species experts in to deal with the situation. We're trying to find a new, suitable area for them to locate the animals. We may have to move them a considerable distance to find proper terrain to sustain their lives. This is proving harder than you might think because of political situations and land uses already in place. It's hard to tell someone that they have to move off their land so some obscure animal can have a home. Where are they supposed to go?" she asked rhetorically.

"I see what you mean. It's hard to care about something else when you're suffering yourself. I can only hope that they see beyond their own situation to the larger picture of species survival. But that is a mighty big hope for people barely surviving themselves. Maybe they will be able to work out some sort of compromise."

"That would be great!" Markus said.

"Well, I need to get back to the task at hand. I'll call you guys later to check in and update you on the current situation," Debra said.

"Catch you later, Debra. We'll pass along your report to the director."

"Thanks, talk to you soon," she said and signed off. After about thirty more minutes Kristen came back into the control room. Natalia gave her Debra's report.

"Next we need an update on the requisition requests from our Squad members around the world and what the status is on filling their needs."

"You've got it, Director," Markus replied. "Here are the lists that we have to date and where we are in filling their needs. It looks like everything is moving along well, we've got equipment coming in from almost the entire North American continent, food and potable water are arriving throughout the affected regions, medical supplies are pretty much everywhere already, and we have Squad members dispersed through every disaster zone. Spare parts to repair machinery are coming in as required, and the rescue efforts are in full swing everywhere. Anything else, Director?"

"No, I think that is good for now," she answered.

CHAPTER SEVENTEEN

The next several days were taken up with the search and rescue efforts of the Squad. There were many sad moments with just as many even more intense happy ones. The final worldwide casualty count would reach into the hundreds of thousands, and the number of injured was well into the millions. More than ten million homes had been destroyed and another thirty million structures damaged to some degree in both of the storms.

The amount of infrastructure damaged was staggering, with some thirty-eight thousand bridges, four million kilometers of roads, railroad track, and canals destroyed, hundreds of thousands of public buildings laid to waste, twenty two international airports knocked out of commission including two of the world's top ten busiest, untold acreages of forest lost, tens of millions of square kilometers of farmland put out of commission for at least several years, hundreds of dams broken and many more needing repair, plus the loss of domesticated animals and of wild birds, animals, and plants.

It was a loss to the planet of unequaled proportion in modern times. The two storms had the destructive force of all the wars fought in the world in the last five hundred years. The best thing, though, was that it could have been a lot worse if the preventive actions and evacuations hadn't happened. Everyone in the Squad could be proud of the job they had done, along with all the governments, volunteer organizations, and just ordinary citizens that pitched in to help the world through the most difficult time in recorded history.

Headquarters had been back on regular shifts for two days now, and everyone was quite well rested again. Silverton had returned to her regular life of sleeping on the director's computer in her office during the day and roaming outside come night. Kristen made plans to visit her family over Christmas and arranged for her friend Karen from England to meet her there. Karen knew her parents, and it would be great if they all got together again.

Meanwhile Markus and Natalia made plans to visit his family in Sweden, and on their next leave of absence they would travel to Brazil. Susan went back to work on upgrading the computers at headquarters, integrating some of the knowledge she just picked up in the real-life situation of the two hurricanes. Paul took a couple of weeks off and returned to Canada for his parents' fiftieth wedding anniversary. Life was returning to normal.

"Sweetheart," Natalia said to Markus when they were alone at her home late one evening, "let's have a World Cup party for the final game on Sunday. We could have everyone over from headquarters for the game, what do you think?"

"Sure, we could do that," Markus answered. "I'll be glad to call everyone and ask them over. Do you have the numbers?"

"Yeah, they're right here. Sure you don't want me to call them?"

"No, I'll be happy to do it. After all, your team won, so it's the least I can do for you."

"Markus, you're such a gracious loser, I love you so much," Natalia said to him and went over and gave him a great big hug and long passionate kiss. She continued, "I don't know if I would be as nice as you're being to me. I only hope that I would, but you know how us Latinos are about life: overly dramatic. But if your country had won I would have been your biggest supporter because I love you so much. Thanks for being mine, darling."

"You know I love you, so what other response could I have?" Markus asked.

"Thanks, honey, I love you." They fell into a long passionate kiss and dropped to the floor.

"I'd better call everyone if we want them to come for our party," he said, slowly disengaging from a very long kiss.

"Okay, but come back here as soon as you're done, sweetheart," she replied.

Markus came back in half an hour and reported, "Well, I was able to reach everyone except for Indira and Paul. Indira wasn't home, but I left a message on her voice mail. Paul is at his parents' fiftieth wedding anniversary. Everyone was excited to come, and they all wanted to bring something, so I told them an appetizer would be great. We can sort of have a pot luck with what they bring, and we can supply the liquid refreshments."

"That sounds great. Plus we can get a couple of trays of food just to make sure we have enough, like maybe artichoke dip, shrimp, and asparagus, and something for dessert."

"Yeah—I know, Ben and Jerry's ice cream and Pepperidge Farm cookies. You know the ones I like, Milanos; they're definitely the best. How does that sound?" he asked as they kissed again.

"That sounds great, honey, and we can get some Korbel to toast the winner, Brazil," she added with a big smile.

"Then it's all set. The game starts at 8:00 p.m. in Moscow, which makes it 10:00 a.m. So I asked everyone over for nine o'clock."

"Do you want to get the food and drinks now?" she asked.

"Sure, and we can grab a bite to eat while we're out. How about Mexican? It's been a while since we went there, and a margarita sounds really great tonight."

"Let's go," she said. "I'm starved, and that sounds really good. I'll drive, my car's out front," and they went to enjoy their first meal out since before they went to level-five at headquarters.

CHAPTER EIGHTEEN

It was a clear crisp September morning in Colorado the Saturday of the World Cup final. The sun rose with just a slight hint of red in the high wispy clouds that were soon to disappear as the sun started its journey into the sky.

Markus and Natalia lay in bed together, holding each other gently, watching the sunrise through their partly drawn shades. The moon was just setting over the mountains out their other window as the planet Venus faded into the morning light. It was the day of their World Cup party, and they were ready. Natalia had high hopes of yet another world championship, but knew that they would face the best team the United States had ever fielded in the game. And she knew that they wanted the win very, very much. Brazil had won the cup more times than any other country, and the United States had never ever even been in the finals before. The U.S. were the underdogs by two goals, but they wanted it badly. It would be a great game, and it would start in only two and a half hours.

"I guess we'd better get up and start getting ready for our guests," Markus said after a long silence.

"I know, but it's so nice just lying next to you with no urgent responsibility," she responded snuggling up closer than ever like two spoons in a drawer.

"You know I love you very much," Markus told her in a quiet whisper into her ear as she lay next to him.

"I know, and I love you too," Natalia answered.

"This has been a great two months, don't you think?" he asked.

"You know it sweetheart."

"Well, we've got only an hour before the first guests arrive, so we'd better get ready. I'll get some orange juice made that we can mix with champagne for some mimosas. We have some tomato juice ready for bloody marys and, of course, coffee for the caffeine set."

"Did you get those Danish at the deli, honey?" she asked. They got up, dressed in their soccer game clothes, and got busy preparing for their party.

"Yes, and they had a special on bagels, so I picked those up too. The store was promoting a new type of potato salad, so I picked up a couple of containers. I think we'll have enough food with what everyone else is bringing also. Do we have enough to drink?"

"I think so. We've got several cases of beer and five or six bottles of wine. We should be all right. That sounds like my cat at the door; she probably wants her food right about now. I'll be back in a moment," she said and went to feed her cat.

There was a knock at the door, and Markus went to answer. It was Susan.

"Hi, Markus. I know I'm a little early, but I'm so excited about the game today that I just couldn't stay away any longer. I'm so ready for the game to start," she exclaimed.

"I know what you mean. Well, we're just finishing getting ready. The TV is in the living room; why don't you go watch the pregame show that's already on."

"Is there anything that I can help with?" she asked.

"Not really," Markus responded, "we're pretty much ready. Natalia is feeding her cat, and she'll be back here in a couple of minutes. Otherwise we're ready to cheer on our team."

"Our team, why Markus you're so darling," Natalia stated as she came back into the room overhearing him talking to Susan.

The doorbell rang, and Markus went to greet the guests. It was Kristen and Indira.

"Welcome, my friends, come on in and make yourselves at home."

"Thanks. Hi, Markus," Kristen answered.

"Who are you two rooting for?" Markus asked.

"Well, I'm for the best team, I can't really pick after Australia lost, and I don't want to upset anyone here, so I just here to watch the game. What about you, Markus?"

"Now that Sweden's lost, I have to be for Brazil. Sorry, Susan, but it's my girlfriend's country."

"Don't worry, Markus, I understand, but you're both going to be disappointed before the day is over," Susan replied.

"I wouldn't count your winnings yet," Natalia piped up. "We're not just going to quit now after all we've gone through to get here."

"Believe me, we don't expect your team to be easy. In fact, we want you to be good so when we beat you, it will mean something."

"Well, we'll see," Markus said. "Anyway, some more people are arriving. I'll answer the door."

"Who is it, honey?" Natalia asked.

"Wow!" Markus said. "How did you get here? I thought you were in India. Give me a hug, cousin."

"I needed to come to headquarters," Helene told him in Swedish, "to file reports on the typhoon, and Kristen told me about your party. We decided it would be fun to surprise you. I'm only here for a few days." They talked to each other for a while longer before going into the room with the rest of guests.

"So you must be Natalia. Markus described you perfectly. I'm very pleased to meet you," Helene said as she entered the room with Natalia.

"And I'm pleased to meet you too. Welcome to Colorado. Is this your first time here?"

"No, I've been here at headquarters several times before, but I must say that I've never spent much time in the U.S. except at headquarters. One of these days I hope to have some time to really visit this part of the world."

"How much time do you have on this trip?" Markus asked.

"Not much at all, I've got to get back to India in a couple of days. The relief effort is in full swing, and I need to assist in the reconstruction effort. We are well along in our highest-priority tasks: rescue of injured, missing, and homeless people; the reconstruction of infrastructure; and helping the general public rebuild homes and businesses. Of course, due to the extent of the damage, it will take years to get everything back to some degree of normalcy. That's why I'm here, to report on the situation, give estimates on what will be needed to rebuild the affected area, and refine the timetable on that reconstruction plan".

"Well, next time you're here," Natalia said, "be sure to make arrangements to stay a little longer, and we'll give you a real tour of this part of North America. There's a lot to see and do here. Great restaurants in Denver to eat at, the Rocky Mountains, and of course Rocky Mountain National Park is a must. Now that wolves, moose, and grizzly bears are back in their original habitats, the Park is a great place to spend the day or for that matter several days.

"Anyway, that will be the next time you're back here. Now we're to watch the World Cup. You know, I'm sure, that this is the first time that every country in the world fielded a team for the competition even though only 32 countries make it to the actual World Cup finals."

"Yeah, I've heard that," Helene said. "That is really great. And now we're down to only two teams left for the final game. Well, I guess I'll be cheering for Brazil since Markus is, not that I don't like the United States. If they do win it will do a lot for the sport in this country. They were the last major nation to embrace soccer, and when they fully do, it will make soccer the first truly international sport. I look forward to that day."

"I know what you mean, but I still hope that we win," Natalia replied.

"No doubt. Sounds like the beginning hype is starting on the television; let's go watch," Helene said.

"Sounds like some more people are at the door Natalia said. "Markus, would you let them in please?"

"Sure honey," he replied. And as he answered the door, "Welcome, Elawa."

"Hey, Markus, how are you doing? Ready for the big game? Where's Natalia?"

"Whoa, that's a lot of questions, but in order: I'm doing great, and yes I'm more than ready for the game to start," Markus answered. "Natalia's in here."

"Sorry, it's just that I'm pretty excited about the game today. Even though Botswana isn't in the final, we did the best we've ever done, and it is sort of fun that our two friends' countries are in the final."

"Yeah, I know what you mean. It's pretty exciting."

"Hey, everyone, Natalia called, "the game is about to start and I want to propose a toast first." She raised her glass. "Here's to all of you, to all of your countries, to the Squad, and to the World Cup."

"Hear, hear," Markus piped in, followed by other utterances of agreement.

"Let me just say one thing before the game starts," Kristen said. "Here's a toast to everyone in the Squad for the effort that you all put out in the last several weeks. The world was under one of the most severe weather attacks in modern history, and you all gave it the best that anyone could have done. Thank you all very much. Now may the best team win," she concluded.

"Does everyone have something to eat and drink, before the game begins?" Natalia asked.

"I could use something to drink. Where would I find that?" Elawa asked.

"Right in here," Markus said, "your choice of tea, Coke, orange juice, beer, wine, champagne, coffee."

"A glass of white wine would be great," she replied. "Thanks"

"You're welcome. Grab a seat, it's about to begin." The starting lineups were being announced, and the crowd was going wild. Many fans from both countries had made the journey to the final game to be there in person, to feel the excitement, to be part of history. It was a great day for sports; according to estimates, almost five billion people, the largest single audience for any single event, would be watching today. It was being broadcast into every nation state on the planet. Brazil was favored by one goal in the latest line on the game. Then the teams lined up, and the game was under way.

EPILOGUE

The twenty-three crucial days of August and September 2038 ended with the United States winning the World Cup with a score of 3–2 in overtime. Recovery continued from the worst weather outbreak in modern history. The legacy of the Squad was to be laid down and still stands today. It was a tough time for the world, and it was a great time for the world. The world showed to just what extent there could be cooperation to combat a common disaster. It was, most say, the Global Disaster Squad's finest hour.

For more information on becoming a member of the Global Disaster
Squad, please contact the GDS at www.globaldisastersquad.com on
the World Wide Web.

Milton Keynes UK
Ingram Content Group UK Ltd.
UKHW050807201123
432900UK00011B/264